Catarina Tivane Nhamposse
Maria Miglino
Edson Liberti

Diet-related morphological and quantitative evaluation of muscle in rats

Catarina Tivane Nhamposse
Maria Miglino
Edson Liberti

Diet-related morphological and quantitative evaluation of muscle in rats

Evaluation of the Gastrocnemius muscle (MG) of wistar rats (rattus norvegicus) subjected to the Mozambique diet

ScienciaScripts

Imprint

Cover image: www.ingimage.com

This book is a translation from the original published under ISBN 978-3-330-99841-4.

Publisher:
Sciencia Scripts
is a trademark of
Dodo Books Indian Ocean Ltd. and OmniScriptum S.R.L publishing group

120 High Road, East Finchley, London, N2 9ED, United Kingdom
Str. Armeneasca 28/1, office 1, Chisinau MD-2012, Republic of Moldova, Europe
Managing Directors: Ieva Konstantinova, Victoria Ursu
info@omniscriptum.com

Printed at: see last page
ISBN: 978-620-8-64429-1

SPECIAL DEDICATION:

Jorge Amílcar Nhamposse, my husband.

"Many waters could not quench this love, nor rivers drown it. Even if someone were to give all the goods of his house for this love, he would be utterly despised (Philippians 1:9)".

My children:

Delightful Catherine, Jocast Samuel, Walter Nhamposse and Wanda Khensha...

For having endured...even when they couldn't, for having understood...even when they couldn't, for having supported...even when they couldn't: the absences, the silence and the distance, especially during the "crucial" moments of the studies;

MY LOVE FOR YOU IS UNCONDITIONAL...I LOVE YOU!

SPECIAL THANKS

My supervisors: Profª Dr Maria Angélica Miglino, who conceived the idea for this project and **Prof Dr Edson Aparecido Liberti,** who took up the cause and carried it out:

I would like to thank you for accepting me as a student and devising this project; I would also like to acknowledge and thank you for all the support you gave me during the difficult times I went through in São Paulo, both administratively and socially. For being understanding and helping me to overcome all the difficulties, especially those that were not academic.

To Professor Maria Angélica Miglino:

- for accepting me as a student, even without knowing me;

- for the intelligent and professional way in which he devised the project and channelled all his personal and material resources into carrying it out;

- for always providing financial support to feed the experimental animals;

- for welcoming me and providing me with many learning opportunities, as well as the moments of relaxation that made it easier for me to join the group of Anatomy postgraduates;

- for having helped me with a number of non-academic issues that were vitally important to my stay in São Paulo;

- for supporting and comforting me in times of sadness, for his willingness to receive and listen to me even during the weekends, and for helping me to overcome all my difficulties.

To Prof Dr Edson Aparecido Liberti and Profª Sílvia de Campos Boldrini *(in memoriam)*

- for welcoming me at a time when they already had in mind a project focused on social issues in Africa, and Prof Edson, even with the pain (certainly greater than ours) of the physical loss of his wife and partner, for having guided me wisely, with the talent and professionalism that characterise him, especially in the "crucial" moments of concluding this project, and even when time was not "our ally", he was able to relieve my apprehension and anxiety with relaxed phrases like "Thesis is not pastry"!

- for having conducted this research wisely, integrating all the LAFACC/VQM colleagues even during my absence, demonstrating once again his charisma as a leader, father and friend.

- for agreeing to turn the VQM into a "nursery" and, together with the VQM colleagues, helping me to look after Wanda Khensha so that I could complete my tasks.

My Parents: Dinis Constantino Tivane *(in memoriam)* and Florinda Manuel Cossa, **Siblings:** Natércia, Costa, Messias, Isaura, Stela, Dinis, **Cousins:** Angelo, Mandinho and Queta, **Sisters-in-law:** Zinha, Daena, Amina, Delucha, Cleta, Vivi, Jotamo, Dias and Tó for their support and encouragement during all these years of "academic life";

My friends and colleagues from the FMVZ Anatomy Department:

- Márcio Nogueira Rodrigues: Thank you, for your friendship, for your strength throughout all this time, for being this companion in academic activities and publications, for your help with the thesis and so many other social helps, for your understanding in difficult times, in short... there are so many THANKS, that almost nothing would fit here. I will always recognise and acknowledge u friendship, no matter how far we will be from each other in a short periodll!

- Paula Fratini: Thank you for being the person you are! Thank you for your strength, a huge THANK YOU for everything you do for me and Khensha!

- Rafael Cardoso: You know how much I value your friendship, and all the help you gave me in the "no team" moments of the thesis! Thank you so much!

...and the others: Érika Fonseca, Marina Brolio, Phelipe Favaron, Carlos Sarmento, Ana Mançanares, André Franciolli, for their friendship, help and good times.

of the LAFACCZ VQM:

- Paulo Henrique, the "in duty" supervisor: Thank you for "guiding" me, especially with the experimental techniques;

- Paulo Henrique, Regina Matos, Cristina Bolina, Joyce Naiara, Alyne Marosti and Lídia: for having conducted the experimental part so wisely even during my absence. I recognise the importance of your valuable contribution and support in this project! Thank you all!

Joice Naiara and Allyne Gonçalves: the baby girls: u friendship will always keep in my heart! Thanks a lot

- the other LAFACC/VQM colleagues: Any Lima, Alyne Gonçalves, Alyne Marosti, Bruna Caixeta, Cristina Bolina, Joice Naiara, Jodonai Barbosa, Paulo Henrique, Regina Matos, Ricardo Eustaquio, Ricardo Bandeira, Diana, Lucilene, Valquíria Mariotti, Natália Camargo, Naiane Klebis, Marcelo Cavali for their help and support in this project, friendship, companionship, strength and, above all, their unconditional support with Wanda Khensha so that I could finish this thesis! Nyabonga ngofo (thank you so much)!

To the following institutions and people for their support:

- Eduardo Mondlane University, for giving me the opportunity to leave the country so that I could continue my studies at doctoral level full-time, and for financing the necessary funds for travelling and settling in São Paulo;

- CAPES, for donating the scholarship for the duration of my studies;

- Mr Maicon Barbosa da Silva and Mrs Rose Elli and Mrs Jaqueline de Santana from the FMVZ/USP Anatomy Department;

- Ladies Daura and Cláudia, and Mr Carlos from the FMVZ/USP Postgraduate Committee;

- Mrs Elza from the FMVZ/USP Library;

- Mr Sérgio from the House of Theses at FMVZ/USP;

- To the technical ladies of the Histology laboratory of the Institute of Sciences of the Anatomy Department of ICB III/USP, Marta, Kelly and Sônia for their patience, dedication and help with histological and ultrastructural techniques.

- To Mrs Rosana, from the ICB Statistics department, for not only doing the statistical analyses "out of hours", but also helping me interpret the final results and allowing me to turn her room into a small "nursery" so that Wanda Khensha could feel "at ease"! Thank you so much!

- Mrs Angelita and her husband, Sandra and her husband, Jacqueline, Adão, Carlos Eduardo for their moral support, strength, help with Wanda Khensha (especially Jacqueline) and moments of relaxation at "Condomínio Angelita".

- To all those who directly or indirectly encouraged and supported me, all those who believed and even those who didn't...., thank you very much!

Ask and it will be given to you...seek and you will find...knock and it will be opened to you....

For everyone who asks receives...

He who seeks finds, and to him who knocks it will be opened...(Matthew 7:7,8)

"The child's mind is virgin and fertile soil. It is therefore not only a necessity, but also an undeniable moral and rational obligation, to contribute to the germination of optimal seeds in the child's small but fertile mental fields, seeds that contain the possibility of manifestation of the resources that man's intelligence needs to emancipate itself from all pressure foreign to its thinking and feeling, and to overcome the difficulties it will face in the course of life...." (*Carlos Bernardo González Pecotche (Raumsol))*

SUMMARY

NHAMPOSSE, C. T. **Morphoquantitative analysis of the effects caused in striated skeletal muscle of Wistar rats *(Rattus norvegicus) subject to* diet utilised in young children of rural Mozambique.** [Structural, Ultrastructural and Morfoquantitative analysis of the effects caused in striated skeletal muscle of Wistar rats (Rattus norvegicus) subject to diet utiled in young children of rural Mozambique]. 2012. 112 f. Thesis (Doctor of Science) - Faculty of Veterinary Medicine and Zootechny, University of São Paulo, São Paulo, 2012.

Mozambique is a country in southern Africa where around 55% of the population lives below the absolute poverty line with less than one meal a day, surviving on donations at great cost. Food insecurity and extremely poor nutrition, especially in children, are factors that lead to levels of Chronic Malnutrition (CD) in children of around 44%. Chronic malnutrition is responsible for one third of deaths in children under five. The aim of this study was to assess the effects of a ration prepared with food produced and consumed by the population of rural Mozambique on the gastrocnemius muscle of wistar rats (Rattus norvegicus) fed this ration. Seventy-five Wistar rats weighing approximately 300g were divided into three groups: Nourished or control (N), Malnourished (D) and Mozambique or experimental group (M), assessed at birth and weaning. The animals were kept under the same housing conditions, temperature, humidity and light, but with different diets depending on the group; Group N with normoprotein feed (20% casein), Group D with malnourished feed (5% casein) and Group M with Mozambique feed. In all groups, body mass was assessed at birth and weaning and the right gastrocnemius muscle of the male chicks was collected at weaning for processing. Serial 10 pm-thick sections were cut using a cryostat and submitted to haematoxylin and eosin, picro-sirius, NADH-tr and transmission electron microscopy techniques. Statistical evaluation of inter-group differences was determined by analysis of variance (ANOVA) and Tukey's test. Significant differences were found between groups N, D and M. The animals in group M showed a large variation in weight and size, which was similar to that of group D; they also showed changes in the shape of the muscle fibres, which had rounded contours and a predominance of type III collagen fibres, as did the malnourished animals. Ultra-structurally, the Mozambique animals showed a misalignment of the sarcomere Z-lines and ruptured myofibrils, a decrease in the cross-sectional area and a lower proportion of glycolytic and glycolytic-oxidative fibres, and a higher percentage and cross-sectional area equal to group D with regard to oxidative fibres.

Keywords: Skeletal muscle. Morphology. NADH-tr. diet. Mozambique.

SUMMARY

CHAPTER 1

INTRODUCTION

1.1 A brief history of the Republic of Mozambique

Mozambique is a sub-Saharan African country with around 18 million inhabitants (Census 2007, Maputo, Mozambique), which belongs to the SADC (Southern African Development Countries) and the CPLP (Community of Portuguese-Speaking Countries). Located in south-west Africa, its entire coastline is bathed by the Indian Ocean (Figure 1).

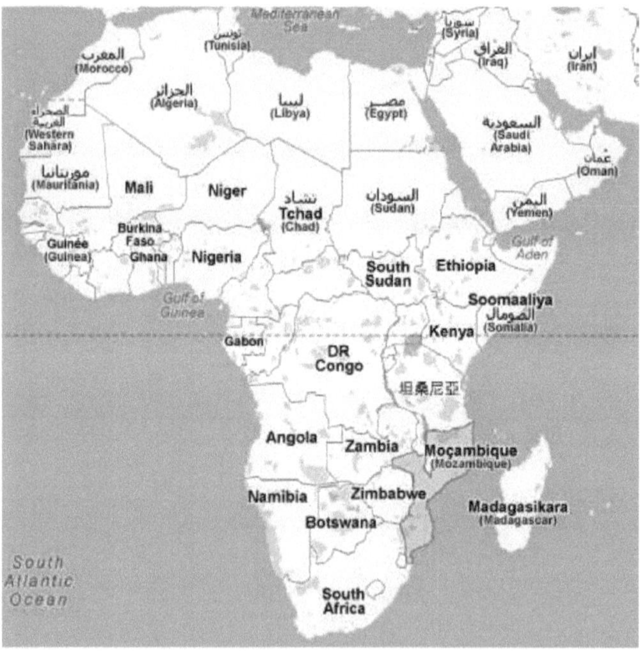

Figura 1- Location of the Republic of Mozambique on the African Continent
Source: Google maps, 2012

Recent figures indicate that 55 per cent of the Mozambican population currently lives below the absolute poverty line (AZZARRI, 2010), surviving very cheaply on donations. Food insecurity and extremely poor nutrition, especially among children, are a fact, with many families getting by on just one meal a day. According to the official publication[1] (Annex A and Table 1) *chronic malnutrition* (CD) in Mozambique is responsible for one third of deaths in children under five. In the country, 1 in 7 children die from CD in the first year of life and 1 in 5 die before reaching the age of five. *CD*, with an average rate of 44% (percentages above 40% are considered "extremely high" by the World Health Organisation), is also a factor responsible for poor productive capacity, absences from the workplace, susceptibility to disease, reduced physical complexion and low intellectual level, which ultimately has a negative impact on the country's ability to achieve the Millennium Development Goals (MDGs) (FINKEL, 2004; SETSAN, 2005; MARTINS, 2008; AZZARRI, 2010).

Reducing child mortality and malnutrition in Mozambique has been a constant challenge, as they continue to claim victims mainly in the country's rural and suburban regions, where basic sanitation is scarce. This is compounded by a lack of knowledge about the value of food, food taboos and other food traditions that are harmful to health (MARTINS, 2008).

Investments in studies to improve the nutritional and health status of human beings as a critical component of a country's economic development are well accepted, particularly in developing countries such as Mozambique. Programmes that promote children's health and nutrition, especially in rural areas, are seen as drivers of a country's equality and growth.

Azzarri (2010) reiterates that improving children's nutritional status is extremely important for the potential growth of countries with a demographic structure like Mozambique's, where most of the population is young (under 15 years old).

As one of the Mozambican government's policies is poverty alleviation, based on the search for solutions to be implemented and developed in rural areas, any project that encourages people to promote self-sufficiency as a form of poverty alleviation is guaranteed to be accepted by the government, as well as by the main international partners.

Haj et al. (1986) mention that an individual's nutritional status is important in determining tissue growth. Rose and Tschirley (2003) describe that the diet of the Mozambican population meets the energy needs of these individuals (since it is rich in peanuts), however, it is believed that a *priori* this diet is not only poor in iron, but also lacks fat and protein in its composition.

There are currently no studies that prove the efficiency of the diet used in rural areas

[1] Republic of Mozambique. 2010. Multisectoral action plan for the reduction of chronic malnutrition in Mozambique 2011-2014 (2020). 111 pages

of Mozambique, which justifies carrying out studies in this area, emphasising the aim of testing whether the introduction of micronutrients in the feed prepared with crops produced in Mozambique increases the diet's efficiency.

In view of the difficulties faced on the African continent and in Mozambique with regard to child nutrition, this project aims to obtain information on the nutritional status of children in suburban and rural areas in Mozambique by investigating the effects exerted by a ration based on locally produced crops/foods on the skeletal muscles of wistar rats.

In this way, it is believed that if the efficiency of the ration formulated from these Mozambican crops is proven, the cultural aspect of the use of locally produced crops will be valued and it will subsequently be possible to propose to the country's Nutrition and Public Health bodies the introduction of this ration in the supplementation of children in rural areas affected by CD.

CHAPTER 2

LITERATURE REVIEW

2.1 DISNUTRITION

The term *malnutrition* (also called *inadequate nutrition)* is generic and can be defined as a clinical condition resulting from the absence or deficiency of any or all nutritional elements (MONTEIRO, 2003). According to the FAO/WHO (2006), malnutrition is defined as insufficient and prolonged food intake associated with low absorption of the food consumed and is applied to calorific, protein or caloric-protein deficiencies.

Although the most frequently reported cause of malnutrition is poor diet, it can be considered a disease caused by an inappropriate hypocaloric and hypoprotein diet (GILLESPIE et al, 1996); Thus, when the nutrients ingested are insufficient for the body to carry out its vital functions, it starts to look for this energy in ATP's, then in glucose from the tissues and blood, in glycogen from the liver and muscles and finally uses the proteins stored in the liver and muscles by breaking them down and transforming them into amino acids. The use of these nutrients to produce energy, when they should have another purpose in the body, causes apathy, prostration, hypoglycaemia, fainting, increased blood acidity, thinning of the skin, loss of muscle mass, weakness of the immune system, iron deficiency anaemia, low absorption of nutrients and can lead to death (VINCENT; RADERMECKER, [19-?]).

Among the various factors that can trigger malnutrition, the *primary* causes are unbalanced diets (a diet that is quantitatively or qualitatively insufficient in calories and nutrients), and *secondary* causes are situations in which there is insufficient food intake due to increased energy needs or another factor not directly related to food, such as anorexia, malabsorption of nutrients by the intestine, excessive loss of nutrients caused by some illness, kidney failure and other aggravating factors (MONTEIRO, 1979; WHO, 1997, 2005).

Malnutrition usually occurs in low-income individuals, in the poor and especially in children in developing countries living in rural and peri-urban areas (WHO, 2002; MONTEIRO, 2003; BOURNE et al., 2007; AZZARRI, 2010).

It has been proven that malnourished people are more prone to diseases, as this weakens the immune system, facilitating the appearance of opportunistic diseases such as diarrhoea, respiratory diseases and others (FINKEL, 2004; WINICHAGOON et al., 2006).

Studies carried out in Brazil report that malnutrition can also be caused by early weaning (children between 0 and 2 years old). These studies point out that these children are usually weaned between two weeks and three months of age, and the food introduced in these cases is not sufficient to meet the nutritional needs of infants, particularly in low-income families. This is coupled with poor sanitary conditions and inadequate hygiene practices, which favours the occurrence of parasites, other infections and diarrhoea. Appetite decreases due to abdominal pain accompanied by fever. The child eats less than normal and consequently has poor growth and development (CARVALHAES; BENÍCIO, 2002; MONTEIRO, 2003).

Malnutrition can also manifest itself in individuals with micronutrient deficiencies (essentially iodine, iron and zinc) and vitamin A deficiencies (HAMPSHIRE et al., 2004; AGUAYO et al., 2004; SETSAN, 2005; HOTZ; GIBSON, 2007; LOW et al., 2007; MANGER et al., 2008; MARTINS, 2008). According to Martins (2008), the lack or insufficiency of micronutrients such as selenium, vitamins C, E and the B complex, lycopene, lithium, omegas 3 and 6, calcium and magnesium also have a major influence on CD.

According to Martins (2008), the indicators of malnutrition are as follows:

1. DC: It is considered by many authors to be the most important and interferes with the child's growth. It reflects the food situation of the family and society. It is measured by the height/height ratio (with a standard deviation of less than 2).

2. *Acute malnutrition: Refers* to situations of food crisis, causing the child to lose weight. With little significance in Mozambique, it is measured by the Weight/Height ratio (with a standard deviation of less than 2).

3. *Weight for age:* This is a mixed indicator for CD and acute. It is not a very specific indicator, but it is the one that is relevant to the Millennium Development Goals in Mozambique (Mozambique Preliminary Report, 2011). It is difficult to interpret, as it has different definitions if it is obtained routinely or during surveys.

1.2 TYPES OF MALNUTRITION

Protein-energy *malnutrition* (PED) is a syndrome that includes a series of symptoms, each of which has a specific cause related to one or more nutrients (such as protein, sludge or calcium). DEP is characterised by the existence of a cellular imbalance between the supply of nutrients and energy on the one hand, and the body's demand to ensure growth and the maintenance of specific functions on the other (JAHOOR et aL, 2008). Its incidence is more frequent in children who are breastfeeding and under 5 years of age, who can also develop iron, folic acid, vitamin C and copper deficiencies if they receive inadequate diets (GIBSON et aL, 2003).

There are three types of *protein-calorie malnutrition* (PCD). In dry PCD, the individual is thin and dehydrated, while wet PCD, also called *kwashiorkor*, is when the individual appears "bloated" due to fluid retention. The name of the disease derives from an African word (Ghana) meaning "the first child's illness because of the second child" and comes from the belief that the first child usually develops *kwashiorkor* when there is no great time difference between the first and second pregnancies. The second child is born and replaces the first child at the mother's breast, so the weaned first child is fed a cereal porridge that has poor nutritional quality compared to breast milk, thus jeopardising its development. Scientifically, it has been shown that protein deficiency in cases of *kwashiorkor* is usually more significant than calorie deficiency, caused by a reduction in plasma proteins, thus reducing the colloidosmotic pressure and, consequently, the osmotic flow through the walls of the capillaries, resulting in fluid retention which translates into generalised oedema, accompanied by skin diseases and hair discolouration (Figures 2 and 3) (NÓBREGA, 1986; VINCENT; RADERMECKER, [19-?]).

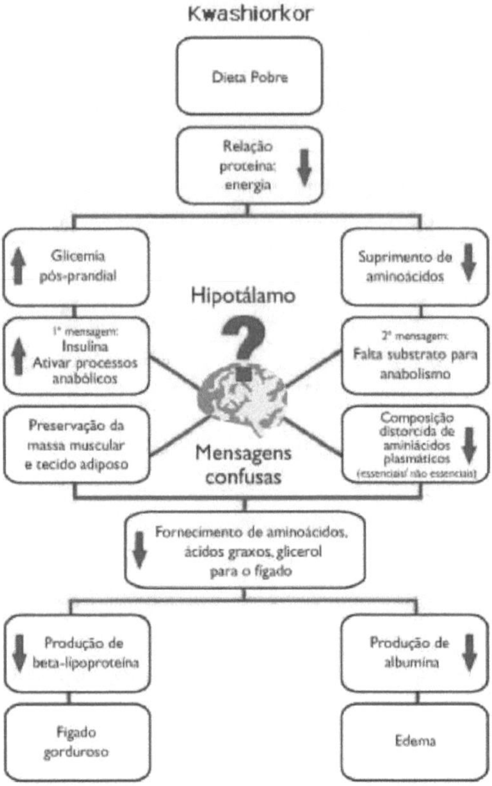

Figura 2- Diagram of the aetiology of Kwashiorkor:

Source: Dr Deo de Freitas, 2012. Kwashiorkor: Metabolic alteration caused by an excess intake of carbohydrates in relation to the amount of protein. The low-protein diet caused a reduction in plasma proteins and a consequent reduction in colloidosmotic pressure with a reduction in osmotic flow through the walls of the capillaries, generating generalised oedema and a gross increase in fat in the liver due to a lack of apolipoproteins, which transport lipids from the liver to other body tissues.

Intermediate CPS, also of the dry type, is known as *marasmus* and results from a typical condition of "hunger". The child with marasmus eats very little food, often because the mother is unable to breastfeed. This child is "extremely" thin due to the loss of muscle mass and body fat. Due to their degree of weakness, they are invariably exposed to any kind of infection. Between *kwashiorkor* and marasmus there is a type of CPS called *kwashiorkor - marasmatic.* Children with this type retain some fluid and have more body fat than those with marasmic (RACELA et al., 1966; JAHOOR et al., 2008).

Kwashiorkor is less common than marasmus and usually occurs as *kwashiorkor - marasmatic.* It is most prevalent in rural areas of Africa, the Caribbean, the Pacific Islands and South-West Asia, where the foods used to wean babies - such as yams, cassava, rice, sweet potatoes and green bananas - are deficient in protein.

Protein deficiency impairs body growth, immunity, healing and the production of enzymes and hormones. Both marasmus and *kwashiorkor* are accompanied by diarrhoea. However, the child recovers more quickly from marasmus than from *kwashiorkor* (JAHOOR et al., 2008).

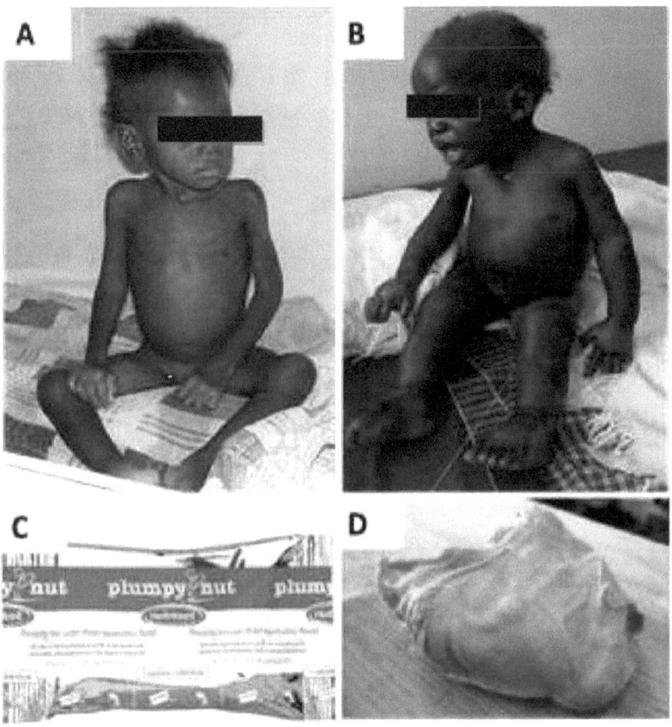

Figure 3- Photograph of children affected by marasmus and *kwashiorkor*

Source: Dr Mandomando, I. Leading Research at ISGlobal. Barcelona Institute of Global Health, 2010. Note in A: extreme thinness due to loss of muscle mass and body fat and B: generalised oedema on the face and limbs, hair discolouration and areas of skin depigmentation with ulcerative lesions all over the body. C and D: ready-to-use peanut-based therapeutic food (plumpy-nut).

Behavioural development can be extremely delayed in severely malnourished children and mental retardation can occur. Up to 40 per cent of children with protein-calorie malnutrition die. Death during the first few days of treatment is usually caused by an electrolyte imbalance, generalised infection, hypothermia or cardiac arrest.

2.3 SOME REFERENCES ON MALNUTRITION AROUND THE WORLD

It is estimated that around 800 million people in the world still cannot meet their basic energy and protein needs, more than 2 billion people lack essential micronutrients, and millions suffer from diseases caused by poor hygiene, food and unbalanced dietary intake. Studies of malnourished children in developing countries show that the growth factors indicated by low weight for age affect around 174 million children under five and 230 million by short stature for age.

Pelletier et al. (1995) showed that in 53 developing countries they tested, the effects of malnutrition on infant mortality (6 to 59 months) were 56%. The individual rate per country varied from 13 to 66 per cent, showing once again the impact that malnutrition has on infant mortality and suggesting a new methodology for estimating the effects of malnutrition on infant mortality in any population where prevalence data are available.

According to the WFP and CEPA, 6.6 million of the 12.2 million deaths among children under five, or 55 per cent of child deaths in developing countries, are due to malnutrition. In addition to the human suffering, this translates into high social and economic costs for the countries affected (MARTINS, 2008).

Poor nutrition during child development can also often be translated into illiteracy. Various studies have shown that the level of

mothers' schooling is a determining factor in the healthy growth of children (SAHN; ALDERMAN, 1997; SETSAN, 2005; MARTINS, 2008).

Children whose parents don't have access to guidance from doctors, nurses, nutritionists, nursing assistants, psychologists, social workers, pre natal care, breastfeeding, weaning, nutritional guidance, vaccination in order to avoid diseases such as diphtheria, whooping cough, tetanus, infantile paralysis, tuberculosis and measles suffer more from the effects of malnutrition than those whose parents had a higher level of education and therefore better living conditions. It can therefore be seen that malnutrition can only be effectively combated through an exchange of integrated social and clinical actions (SAHN; ALDERMAN, 1997; AZZARRI, 2010).

Data on the long-term effects of child malnutrition are still inconsistent. Some studies show that when these children are treated in time, the liver, muscles and immune system recover completely (OLDFORDS; SOURANDER, 1986; JAHOOR et al., 2008). However, in some cases, there is impaired absorption of nutrients by the intestine and varying degrees of mental impairment, which are directly related to the length of time the child has been undernourished, the severity of the illness and the age at which the illness began (MARTINS, 2008; AZZARI, 2010).

2.4 STUDIES ON MALNUTRITION ON THE AFRICAN CONTINENT AND IN MOZAMBIQUE

In Africa, several studies have been carried out on the availability of essential micronutrients and vitamins that exist or have been introduced into people's diets.

Ferguson et al. (1993) investigated the level of zinc in children in rural areas of south-west Ghana and southern Malawi by determining the concentration of the mineral in the children's hair, as well as anthropometric indices of growth and body composition. The authors observed that Malawian children (on a cereal-based diet) consumed little fat and higher amounts of energy, protein, zinc, calcium and phytates than Ghanaian children (accustomed to a diet based on cereals and starch, cassava and other vegetables). The authors concluded that although children in both countries consumed inadequate levels of zinc, the prevalence of children with low growth was higher in Malawi than in Ghana.

In 1997, Sahn and Alderman evaluated the effects of malnutrition on children in rural and peri-urban areas of Maputo (capital of Mozambique) aged 0 to 2 years (group 1) and 2 years and one month to 6 years (group 2) using anthropometric measurements. These authors concluded that increasing parental income improved nutrition, and that transferring income to other household expenses had a slight impact on children's anthropometric nutrition indicators compared to other sources. However, the impact of income was only more noticeable in children over two years old. On the other hand, the mother's education, access to improved latrines and the mother's height were considered determinants of height for age. The authors agreed that programmes to improve children's weight, accompanied by nutritional education and preventive care, are considered "key tools" for improving anthropometric indicators of nutrition.

In a collaboration between the University of Porto (Portugal) and the Higher Institute of Science and Technology of Mozambique (ISCTEM), Lunet et al. (2002) carried out a study of children under two years of age to determine their malnutrition parameters. This study was carried out at the Polana-Caniço Health Centre, on the outskirts of Maputo City, Mozambique. The prevalence of low length and weight for age, and low weight for length and their association with socio-demographic factors, maternal characteristics and the child's diet were considered. These authors concluded that malnutrition was expressed through the low height/age ratio. On the other hand, low height/age or low weight/age were directly related to poor housing, home births, low schooling and low maternal height.

Rose and Tschirley (2003) evaluated dietary improvement through monitoring home feeding in the provinces of Nampula and Cabo Delgado in Mozambique; and in 2003, Gibson et al. contributed strategies for the adequacy of micronutrients in diets based on maize and low in animal protein, for children aged 30 to 90 months in rural areas of southern Malawi. Both studies demonstrated the importance of dietary monitoring in promoting daily nutrient requirements, so that these produce good growth, health and intellectual development in children.

Anthropometric indices (weight, height, head and chest circumference, skinfolds, etc.) as indicators of the nutritional status of children and adolescents aged between 6 and 18 were assessed by Prista et aL (2004) in the city of Maputo, Mozambique. However, these researchers applied their studies to assessing the physical activities of these individuals, and gave no information about the effects of malnutrition on their development.

Hotz and Gibson's (2007) experiments with supplementation diets, based on alternative foods, brought satisfactory results in Malawi. Using homemade techniques, the authors used a variety of traditional food processing strategies with the aim of maximising the availability of micronutrients in the vegetables consumed in children's diets, combining a series of techniques that ranged from thermal and mechanical processing, soaking, fermentation and germination to the inclusion of micronutrient-rich foods and animal protein in the diet. As a result, they observed improvements in haemoglobin levels, body mass and plasma retinol levels, and consequently a lower incidence of infections.

Recent information shows that in the province of Zambézia (located in central Mozambique), a region often plagued by drought and high levels of child malnutrition, a group of American researchers (LOW et aL, 2007) in collaboration with other Mozambicans designed the first Vitamin A supplementation programme for children in the country and on the African continent, by introducing orange-fleshed sweet potatoes into their diet. The aim of this project was to combat Vitamin A deficiency, which affects around 71 per cent of children aged between 6 and 59 months, mainly in rural areas of Mozambique. Although the Mozambican government has made recent efforts to make free Vitamin A available to all children under the age of 5, it has been found that a Vitamin A supplementation programme alone does not solve the problem of Vitamin A deficiency in children. Therefore, the strategy used by the researchers to introduce and promote foods enriched with vitamin A (such as the orange-fleshed sweet potato mentioned above) at community level indicates that the study was successful, since the orange-fleshed sweet potato was well accepted by the children, providing them with 90 per cent vitamin A. The results confirmed that this sweet potato was regularly consumed by the children during the time of the study, even during the dry season, with more than 50 per cent of the children consuming it around 3 or more times a day.

Also in 2007, a study was carried out in South Africa emphasising the importance and motivation of following the FAO/WHO recommendations (1998) and introducing the "Food Based Dietary Guidelines (FBDG)" nutrition education strategy based on local development directly for adults and schoolchildren (BOURNE et aL, 2007). Within this framework, the country's national nutrition programme has created eight areas with special focus on: counselling, support and treatment of diseases caused by nutritional deficiencies; growth promotion and monitoring (GMP); control of malnutrition by introducing micronutrients into the diet; promotion, protection and support of breastfeeding; contribution of food security in families; nutritional interventions in HIV-infected children and the promotion of nutrition, education and support. Progress towards these goals has included a "child-friendly hospital" initiative, "mandatory" enrichment of maize and wheat flour with multiple micronutrients such as folic acid; national coverage with vitamin A supplementation and mandatory salt iodisation by legislation.

Much more recently, in April 2010, the first therapeutic food factory based on peanuts, powdered milk, oil, sugar, vitamins, minerals and water (APTU-"plumpy-nut") was opened in Mozambique (see Figure 3 above) for the treatment of acute malnutrition, which until now has been done through hospital treatment with essential medicines and specific therapeutic milks (called F-75 and F-100,) involving hospitalisation of the child for around 3 to 4 weeks, in the company of the mother. Plumpy-nut has been used successfully in other African countries such as Niger, the Democratic Republic of Congo, Ethiopia, Madagascar, Malawi, Tanzania, and others such as the Dominican Republic and India (www.unicef.org/mozambique).

2.5 GENERAL INFORMATION ABOUT DC IN MOZAMBIQUE

In 1997, Shrimpton estimated that CD in Mozambique affected 36% of the population, resulting in a loss of productivity of around 2 to 3% of GDP, which represented a high cost to the country. Using this same basis of calculation, taking into account current malnutrition data and based on the 2006 GDP, the same author points out that low productivity cost the country between 205 and 273 million dollars in 2006. As a result of malnutrition, from 2004 to 2008, there were 208,000 child deaths due to protein-calorie malnutrition, 163,000 child deaths due to Vitamin A deficiency and 1,327,000 newborns with mental retardation due to sludge deficiency (MISAU, 2002; MARTINS, 2008).

According to Martins (2008), in 1997 the CD was 36% (18 times higher than normal). The provinces with the highest rates were Cabo Delgado (57%), Niassa (55%) and Tete (46%). In 2001 it rose to 44 per cent, and the highest rate was in Zambézia Province (62 per cent). In 2003 it was 41 per cent. The worst provinces were Cabo Delgado, Zambézia, Niassa and Tete, all in the centre and north of the country. The current rates (2011) are shown in Annex A, Table 1.

Recent studies report 44 per cent malnutrition in the country (AZZARI, 2010). The worst rates are in children of illiterate mothers, in children who do not live with their mothers and in rural areas (MARTINS, 2008). On the other hand, Azzarri (2010) reported that in addition to these, the characteristics of the mother (mainly height and fertility) and family income, plus the high dependence of malnutrition on the seasonal nature of the country and food insecurity are other factors that may be correlated with CD in Mozambique.

2.5.1 Causes of malnutrition in Mozambique

Martins (2008) and Azzarri (2010) point to the main causes of malnutrition in Mozambique:

(i) Poor access to food, due to food shortages (insufficient production and/or import capacity). This is due to the low power of the commercialisation network, instability in food supply or the economic inability of many households to buy food;

(ii) Lack of knowledge of the value of food, associated with food taboos and other food traditions that are harmful to health and common in rural areas;

(iii) malnourished children who survive, 12 to 15 years later will have a weak physical complexion, low intellectual level, low productive capacity, many days of work lost, low immunity and consequently greater susceptibility and high levels of mortality.

Combating malnutrition in Mozambique is a government priority (SETSAN, 2005). Although the Ready-to-Use Therapeutic Food (APTU/"plumpy-nut") factory has recently been inaugurated in the city of Beira in the centre of the country, in an effort to intensify outpatient treatment interventions for severe acute malnutrition (www.unicef.org/mozambique), recent studies show that there is almost no information on the nutritional requirements for preparing a balanced diet using local crops for children with CD (BANDEIRA et al., 2001; ROSE; TSCHIRLEY, 2003; PRISTA et al., 2003; HAMPSHIRE et al., 2004; ROSENTHAL et al., 2004; LOW et al., 2007). This is partly due to the fact that important data on the energy, protein, lipid and macro and micronutrient content of these crops is not adequately publicised.

Since the diversification of livelihoods is necessary for poverty alleviation in Mozambique, there is no doubt that the adoption of a balanced ration using sustainable technology in the processing of local crops will make a valuable contribution to improving the nutritional status of rural populations, particularly children.

Alternative feeding to supplement malnourished or starving children has been used successfully in many countries. In Brazil, the idea began to be disseminated by the "Pastoral da Criança" and consists of making the most of food, avoiding waste and looking for new sources of nutrients. The best-known diet is Multiple Flour, made with bran (rice and wheat), leafy greens (cassava, sweet potato, pumpkin) and seeds (sunflower, watermelon, etc.), which has been proven to be a rich source of nutrients (http://www.pastoraldacriança.org.br).

2.6 MALNUTRITION AND SKELETAL MUSCLE

2.6.1 Muscle tissue

Muscle is the only organ in the body that converts chemical energy into mechanical energy and thus performs the most varied functions; therefore, an inadequate supply of calories and proteins will have the direct consequence of morpho-functional alterations to this organ (NASCIMENTO et aL, 1990).

Muscle tissue is made up of elongated cells arranged in parallel, rich in cytoplasmic filaments of contractile proteins that generate the force needed to contract this tissue through the energy contained in ATP molecules, and consequently becomes responsible for the movement of the body and changes in the size and shape of internal organs (JUNQUEIRA; CARNEIRO, 2004; ROSS; PAWLINA, 2006). Muscle cells originate from the mesoderm and differentiate by synthesising filamentous proteins at the same time as the cells elongate.

There is some controversy regarding the division of muscle tissue types. Authors such as Ross and Pawlina (2006) describe only two types of muscle tissue in the body: striated muscle (skeletal, visceral and cardiac) and smooth muscle, while others such as Junqueira and Carneiro (2004) and Standring et aL (2008) describe three types of muscle tissue according to their morphological and functional characteristics:

(i) Skeletal striated muscle: formed by bundles of cylindrical cells arranged in parallel which, under optical microscopy, show transverse striations; the nuclei of its fibres are located on the periphery of the cell, below the plasma membrane.

(ii) Cardiac striated muscle: cardiac muscle cells and their fibres are transversely striated and the nuclei are centrally striated. In addition,

show intercalated discs, which are strongly coloured transverse bands that cross the fibres, giving the appearance of "steps".

(iii) smooth muscle: whose cells do not show transverse striations, are elongated and the nuclei of the cells in cross-section have a circular appearance, while in longitudinal section they follow the elongated shape of their fibres and appear elongated and also with tapered ends.

2.6.1.1 Development of the skeletal striated muscle

The muscle cells (also called muscle fibres or myocytes) originate from the reserve of pre-myoblastic cells in the demnatomiotome of the mature somite in response to molecular signals from neighbouring tissues and begin to differentiate into myoblasts between the fourth and fifth week of gestation. At six weeks, the cells have already migrated from the dermatomyotome compartment to form the myotome in the centre of the somite (Figure 4A). The myoblasts that form the skeletal muscles derive from the mesoderm in the region of the somite myotomes. The limb muscles develop from the limb buds in the embryo (Figure 4B). The precursor myogenic cells migrate to the limb buds and undergo an epithelial-mesenchymal transformation. The elongation of the nuclei and cell bodies of the mesenchymal cells as they differentiate into myoblasts is the first indication of myogenesis. Next, the primordial cells fuse and form cylindrical, elongated and multinucleated structures called myotubes. All of these transformations are preceded by the activation and expression of the MyoD genes of muscle-specific transcription factors such as MyoD, myogenin, MYf-5 and MRF4 in myogenic precursor cells (MOORE; PERSAUD, 2008).

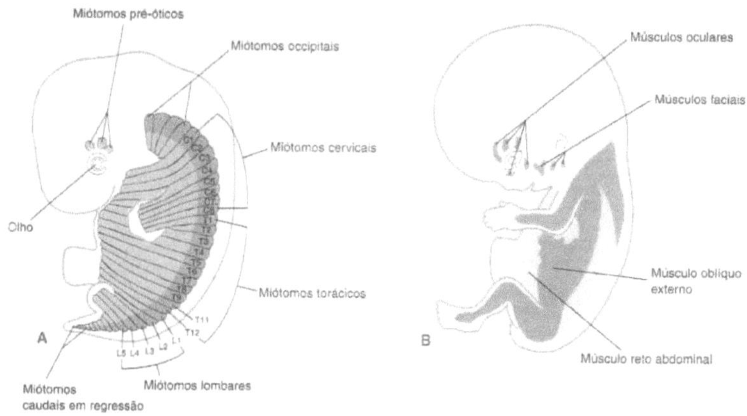

Figure 4- Embryogenesis of skeletal muscles.

Source: Moore and Persaud, 2008. A: 6-week-old embryo showing the myotome regions of the somites that give rise to the skeletal muscles. In B: 8-week-old embryo where the developing muscles of the trunk and limbs can already be seen.

Muscle growth during development is ensured by the continuous fusion of myoblasts and myotubes. Myofilaments develop in the cytoplasm of myotubes during and after myoblast fusion. Myofibres and other organelles then develop. Because of their elongated nature, muscle cells are called *muscle fibres.* The myotubes differentiate and are covered by the outer laminae that separate them from the surrounding connective tissue. Fibroblasts give rise to the *perimysium* and *epimysium.* The *endomysium* is formed by the outer lamina and reticular fibres. Almost all skeletal muscles develop before birth and the rest are formed by the end of the first year of life. The increase in muscle size after the first year of life results from an increase in the diameter of the muscle fibres due to the formation of more myofilaments. Muscles grow in length and thickness in proportion to the growth of the skeleton (MOORE; PERSAUD, 2008).

In muscle cells, filaments of actin, myosin and their proteins fill the interior of the cell. These filaments are grouped in a single direction in such a way that molecular interactions are transmitted in a linear contraction. The most important property of these cells is their ability to change their shape. The grouping of contractile muscle cells forms the muscle, which in turn converts chemical energy into mechanical energy that allows the body to carry out a wide variety of functions such as locomotion, respiration, circulation, etc. In humans, muscle tissue makes up 40 to 50 per cent of the body's mass (STANDRING et al, 2008).

2.6.1.2 Skeletal striated muscle

Skeletal striated muscle is made up of bundles of very long (up to 30cm), cylindrical, multinucleated cells containing many filaments, the *myofibrils*, which represent the structural and functional subunit of the muscle fibre. These bundles are surrounded by a layer of fibrous connective tissue: the *epimysium*, which covers the entire muscle. The epimysium in turn emits septa into the muscle, separating it into bundles or groups of muscle fibres. These septa are called *perimysium* and each muscle fibre is individually surrounded by the *endomysium*, which is formed by the basal lamina of the muscle fibre in association with reticular fibres (Figure 5).

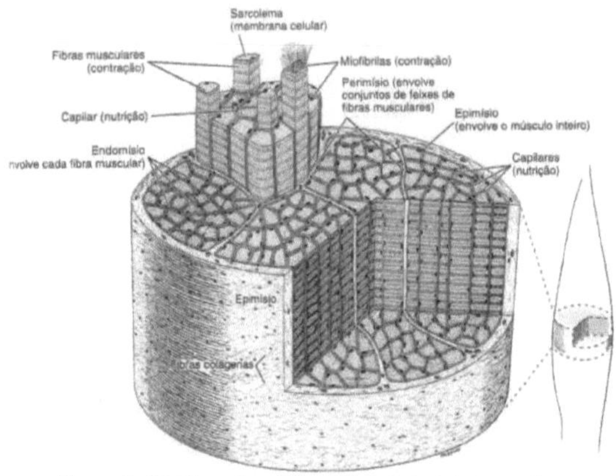

Figure 5- Stratigraphy of the striated skeletal muscle
Source: Junqueira and Carneiro, 2004. The fibrous connective tissue (epimysium) covering the entire muscle fibre can be seen, sending out septa (perimysium) that surround the fibres as a whole, which are individually wrapped one by one by the endomysium.

The diameter of skeletal striated muscle fibres varies between 10 and 100pm. In skeletal muscle fibres, the nuclei are located on the periphery of the fibres close to the sarcolemma. This location of the nuclei is typical of skeletal muscle and helps differentiate it from cardiac muscle, because although both have transverse striations, in cardiac muscle the nuclei are located centrally (JUNQUEIRA; CARNEIRO, 2004; ROSS; PAWLINA, 2006; STANDRING et aL, 2008).

2.6.1.3 Types of muscle fibres

Ranvier[2] (1873 apud EDGERTON; SIMPSON, 1969) first described red and white fibres, emphasising that they were morphologically different and that the two types of fibre could be mixed in the same muscle. Ogata (1958) classified muscle fibres as red, intermediate and white. Brooke and Kaiser (1970) recognised three main types of muscle fibres, called type I, HA and IIB fibres, and a third subtype of type II fibres, called IIC fibres, according to

---

[2] RANVIER, L. Propriétés et structures différentes des muscles rouges et des muscles blancs, chez les lapins et chez les rales. *Acad. Sei. v.* 77, p.1030, 1873.

the reaction pattern for ATPase activity in the globular part of myosin.

Peter, Barnárd and Edgerton, (1972); Simoneau and Bouchard (1995); Pette and Staron (2001) and D' Antona et aL (2006) defined the following types of muscle fibres:

1 - Slow *twitch fibres:* **Type I** (slow-twitch fibres), dependent on oxidative metabolism (SO - slow oxidative).

2 - Fast *twitch fibres:* **Type HA** (fast-twitch fibres), dependent on oxidative and glycolytic metabolism (FOG - fast oxidative and glycolytic).

3 - *Fast twitch fibres:* **Type IIB** (fast-twitch fibres), dependent on glycolytic metabolism (FG - fast-twitch glycolytic).

Similarly to these authors, much more recently, Ross and Pawlina (2006) described three types of skeletal muscle fibres, which differ in terms of diameter and colour. These are *red, white and intermediate fibres.* However, the differences in colour of these fibres are not visualised in histological sections stained with Haematoxylin-Eosin (HE). For this purpose, it is necessary to process the samples with histochemical reactions based on the activity of oxidative enzymes, more specifically the succinic dehydrogenase and nicotinamide adenine dinucleotide-tetrazolium (NADH- tr) reactions, which reveal the three types of skeletal muscle fibres defined above:

- *Type I or slow oxidative fibres:* appear red in fresh specimens, contain many mitochondria and myoglobin, as well as cytochrome complexes in large quantities. The high level of oxidative enzymes in the mitochondria is demonstrated by a strong histochemical staining reaction. They are deeply stained in a darker tone by the NADH-tr reaction. These fibres have a *slow and prolonged contraction and are very resistant to fatigue.*

- *Type IIA or fast oxidative glycolytic fibres:* these are also called "intermediate" fibres. They have an intermediate colour, medium size, many mitochondria and a high myoglobin content. These fibres, unlike type I fibres, contain a large amount of glycogen and carry out anaerobic glycolysis. They are capable of *rapid contractions and are resistant to fatigue.*

- *Type IIB or fast glycolytic fibres:* these fibres contain less myoglobin and fewer mitochondria than type I and Ha fibres. They have a low level of oxidative enzymes and high anaerobic enzyme activity, storing a significant amount of glycogen and thus producing lactic acid, which is responsible for the fibres' rapid fatigue. They are *fast twitch* fibres *prone to fatigue* and generate high maximum muscle tension, which is why they are adapted to fine and precise movements.

In the lower limb, both types of muscle fibres can be found in all muscles, with only the portion between these two types of fibres differing, which characterises the contractile pattern of the skeletal striated muscle. However, muscles in more superficial regions have a greater number of fast twitch fibres than muscles in deeper regions, such as those adjacent to bone structures (HAJ et aL, 1986; HUGHES; BI_AU, 1992).

According to Brooke, Williamson and Kaiser (1971), the differentiation of fibre types begins gradually from the 22nd week of foetal life, after innervation has already been established. These authors describe IIC fibres as undifferentiated and precursors of type I fibres, as well as type HA and IIB fibres. After 20 weeks of gestation, type I fibres begin to appear. After 30 weeks of gestation, type IIA and IIB fibres appear and it is possible to see their typical mosaic pattern, when the fibres are already capable of expressing fast twitch myosin heavy chain (MHC) isoforms. The authors also mentioned that, at birth, the differentiation process is not complete and that there are around 15 to 20% of undifferentiated IIC fibres and that the proportion of HA fibres is even greater than that of IIB fibres (FARKAS-BARGETON et al, 1977; COLLING-SALTIN, 1978), while in adults there are only one to two percent of IIC fibres (SARNAT, 1982; ELDER; KAKULAS, 1993; CARPENTER; KARPATTI, 2001). On the other hand, Dubowitz (1985) considers IIC-type fibres to be rare in normal human muscle; however, for this author, these fibres can be present in developing muscle and can also occur in pathological circumstances.

2.6.1.4 The gastrocnemius muscle of man and rat

The gastrocnemius muscle is made up of two heads; the medial head is the largest and begins in a depression in the upper and posterior part of the medial condyle of the femur. The lateral head begins on the lateral condyle. Both heads can also emerge from underlying areas of the knee joint capsule. The tendon junctions expand to cover the posterior surface of each head with an aponeurosis from the anterior surface of where the muscle starts. The muscular part extends almost halfway down the back of the leg. The fibres of the medial head extend more distally than those of the lateral head. As the muscle descends, the muscle fibres begin to insert into a wide aponeurosis that develops on its anterior surface. Until this point, the muscle masses of the two heads remained separate. The aponeurosis gradually narrows and receives the tendon of the soleus muscle on its deep surface to form the calcaneal tendon (STANDRING et al., 2008).

There are two muscle groups in the posterior part of the rat's pelvic limb. The superficial group is made up of the triceps sural muscle, formed by the gastrocnemius and soleus muscles, which originate independently and end in a single tendon, and the plantar muscle. The deep group is made up of the popliteus, flexor hallucis longus, flexor digitorum longus and tibialis posterior muscles (GREENE, 1959).

The gastrocnemius muscle (MGc) has a predominance of type II fibres. However, in the medial head of the rat's gastrocnemius muscle there are three types of fibres, distributed in a mosaic pattern, similar to that found in human muscle (BROOKE; KAISER, 1970; SHER; CARDASIS, 1976; BANKER; ENGEL, 1994). Hamalainen and Pette (1993) described type IID fibres in the gastrocnemius muscle of rats visualised through the histochemical reaction of ATPase. Staron et al. (1999) observed type I, HA, IIAD, IID, IIDB, IIB fibres in the superficial part of the gastrocnemius muscle of rats, and type I, IC, IIC, HA, IIAD, IID, IIDB, IIB fibres in the deep part, using ATPase histochemical reactions with pre-incubation at different pHs.

In recent studies, Oliveira (2006) determined through histochemical reactions of NADH-tr and ATPase that factors such as nutrition and thermal injury caused changes in the percentage of muscle fibre types in malnourished Wistar rats subjected to thermal injury by scalding. In fact, this author described that the area of the profile of type I, II, HA and IIB fibres was smaller in the malnourished animals, since in the animals with thermal injury this area was smaller only for HA fibres at 4 days after thermal injury and IIB fibres at 4 and 14 days after thermal injury.

Mayer (2008) visualised an interconversion of type IIB fibres into HA through the histochemical reactions of ATPase and NADH-tr in the tibialis anterior and soleus muscles of young rats submitted to immobilisation of the talocrural joint.

Alves (2010) described through histochemical reactions of NADH-tr in the medial part of the gastrocnemius muscle of wistar rats submitted to tenotomy and tenorrhaphy, that in the group tenotomised at 21 post-surgery there was a reduction in the cross-sectional area of the Glycolytic fibres (type IIB) in the superficial region, while the Oxidative fibres (type I) in the deep region remained similar to the Control group, suggesting that the changes in the Oxidative fibres occurred in the first 14 days after the injury.

The gastrocnemius muscle was chosen for this study due to its superficial location, which facilitates its rapid dissection. In addition, several authors (Brooke; Kaiser, 1970; Brooke; Williamson; Kaiser, 1971) have confirmed that the fibre types of the rat's gastrocnemius muscle are equivalent to the fibre types found in the human gastrocnemius muscle, which will help us extrapolate the results of this experiment to humans.

2.6.2 Experimental models of malnutrition

Since child malnutrition is considered a serious public health problem, affecting more than a third of the world's child population, many studies have been carried out to assess the effects of various forms of malnutrition on different tissues, organs and organ systems. As long ago as 1967, Zeman demonstrated in his study that the body weight and survival rate of the offspring of females fed a hypoprotein diet (6% casein) were lower than those of the offspring of females fed a diet containing 24% casein (hyperprotein) during pregnancy. Knowing that nutritional status is important in determining tissue growth (HAJ et al, 1986), experimental studies have already been carried out showing that a state of malnutrition can be reproduced in laboratory animals using different methods, such as restricting the food supply (HEDLEY-WHITE; MEUSER, 1971; HEDLEY-WHITE; MEUSER, 1973; SIMA, 1974; BEDI; WARREN, 1983; CONBOY et al, 1987; CORNBLATH; BROWN, 1988), the use of a low-protein diet (ROY et al., 1972; OLDFORDS; PERSSON, 1982; CONRADI et al., 1985,1989; RANA et al., 1991) or a high-protein diet (BRANDÃO, 1998; CASTELUCCI, 1999), or the concomitant use of restricted food supply and litter expansion (WINICK; NOBLE, 1966; GOPINATH et al., 1983).

Particularly in the striated skeletal muscle, among the morphological changes that malnutrition can cause in the offspring of various animal species, several authors are unanimous in stating that the most obvious is a reduction in the diameter and volume of muscle fibres, as well as a delay in the maturation and development of muscle fibres (ROY et aL, 1972; MAKI; CAMPOS, 1975; GOPINATH et aL, 1983; HAJ et aL, 1986; NASCIMENTO et aL, 1990). Williams and Hughes (1978) showed in their studies on the triceps sural muscle of rats subjected to neonatal malnutrition that muscle mass growth occurs unsatisfactorily, since the width of the muscle was smaller in the malnourished animals than in the control group at the same age. In previous studies, Timson and Dudenhoeffer (1985) subjected young rats to severe protein restriction and observed that in the soleus and tibialis anterior muscles there was a decrease in the rate of muscle growth as well as in the size and number of muscle fibres.

Likewise, Montgomery (1962), Enesco and Puddy (1964) and Oliveira et aL (1999) agree that the effects of early malnutrition on muscle development may be permanent, similarly to the results of previous studies which have shown that in states of malnutrition there is an irreversible decrease in the number of muscle fibres (GOPINATH et aL, 1983; MONTGOMERY, 1962; ROY et aL, 1972), with a decrease in type HA (glycolytic-oxidative) fibres in relation to those of subtype IIB (fast glycolytic) (OLDFORDS; SOURANDER, 1986;

SIECK; LEWIS; BLANCO, 1989; OLIVEIRA et aL, 1999).

Racela et al (1966) experimentally assessed monkeys kept in a clinical condition of kwashiorkor under transmission electron microscopy and reported muscle atrophy with a reduction in calibre and loss of myofilaments, enlarged and irregular mitochondria. Mitochondrial loss has also been described in the muscle of rats subjected to protein malnutrition (HANSEN-SMITH; MAKSUD; VAN HORN, 1977; OLDFORS; SOURANDER, 1983) as well as a reduction in the activity of the succinate dehydrogenase (SDH) enzyme used for mitochondrial analysis of muscle fibres (OLDFORS; SOURANDER, 1986) and a reduction in the activity of the NADH muscle enzyme (NASCIMENTO, 1990).

Considering the information in the literature regarding the morphological changes that skeletal muscle undergoes in situations of clinical malnutrition, this study aims to quantitatively and qualitatively assess these changes in the gastrocnemius muscle of wistar rats subjected to a diet used in rural and suburban areas of Mozambique in order to correlate the effects that this diet may have on the muscular development of malnourished children from these areas in the country.

Thus, in the challenges to create programmes that help alleviate the poverty of rural populations in Mozambique, the Eduardo Mondlane University (UEM) has the important mission of training professionals in multidisciplinary areas of knowledge who contribute ideas aimed at improving the lives of rural populations and sustainable development as a way of eradicating absolute poverty in the country.

Thus, this project is a unique opportunity for UEM, as the largest and most recognised public higher education institution in the country, in its role of disseminating research results that support rural communities and low-income populations through inter-institutional exchange and the training of its staff, teachers and students.

The aim of this research is initially to enable a UEM lecturer to obtain a doctorate so that she can then provide specific training in her home country for technical staff directly involved in feed/diet production using locally produced cultures in the Anatomy and Physiology laboratories of the UEM Veterinary Faculty, and in the areas related to nutrition and food safety at the Ministry of Health (MISAU) in Maputo, Mozambique.

This is an innovative proposal, as it aims to investigate the effects on the components of the gastrocnemius muscle of wistar rats fed a diet made from crops grown locally in Mozambique. Comparative morphological aspects of the muscle development of these animals were assessed in relation to animals of the same age (21 days) that were malnourished (given a hypoprotein diet with 5% casein) and nourished (given a normoprotein diet with 20% casein).

CHAPTER 3

OBJECTIVES

3.1 GENERAL OBJECTIVES

Considering the existence of a correlation between muscle development and diet, and knowing that malnutrition negatively affects this development, the aim is to evaluate the morpho-quantitative aspects of gastrocnemius muscle (MGc) development in wistar rats *(Rattus norvegicus)* fed a diet based on locally produced crops in Mozambique, in order to assess the degree of impairment of the effects of this diet on the Skeletal Muscles of children from rural and peri-urban regions of Mozambique.

3.2 SPECIFIC OBJECTIVES

To evaluate the effects of the Mozambique diet on the following parameters in the gastrocnemius muscle of wistar rats using light and transmission electron microscopy techniques:

a- Qualitative

- the structure and ultrastructure of muscle fibres and their envelopes;
- the collagen constituent of muscle wraps, by typing type I and type III collagen fibres;
- the types of muscle fibres through histochemical analysis (oxidative, glycolytic-oxidative and glycolytic);

b- Quantitative;

- fibre density (oxidative, glycolytic-oxidative and glycolytic);
- the cross-sectional area (ASTf) of the muscle fibres.

CHAPTER 4

MATERIAL AND METHOD

All the experiments were carried out in accordance with the principles of laboratory animal care in accordance with Law No. 6638 of the Federative Republic of Brazil, which deals with norms for the didactic-scientific practice of animal vivisection (BRASIL, 1979), under bioethics protocol number 2223/2011 approved by the Ethics Committee for the use of animals at the Faculty of Veterinary Medicine and Zootechny of the University of São Paulo (FMVZ/USP).

4.1 METHODOLOGY

Young male and female wistar rats *(Rattus norvegicus)* weighing between 280 and 320 grams were mated for a period of ten days, during which time the specific diets for each group were provided without restriction;

- Nourished Group (N): this group consisted of animals that were fed during the mating and gestation period (females) with a normoprotein AIN-93G[2] ration (20% casein) and water *ad libitum.* The animals were fed this diet until weaning and sacrifice of the pups (21 days), which were used in the experiment as the CONTROL GROUP.

- Malnourished Group (D): made up of animals fed during the mating and gestation period (females) with AIN-93G1 (5% casein) hypoprotein feed and water *ad libitum.* The animals were fed this diet until weaning and sacrifice of the pups (21 days), which were used in the experiment as EXPERIMENTAL GROUP 1.

- Mozambique group (M): the animals in this group were fed Mozambique feed and water *ad libitum* during the mating and gestation period (females). The animals were fed this feed until weaning and sacrifice of the pups (21 days), which were used in the experiment as EXPERIMENTAL GROUP 2;

4.1.1 Mating, pregnancy and obtaining litters

This study used 75 Wistar *rats*. The experimental animals were obtained from matings between sires/breeders subjected to the three diets described in the diagram below: Flowchart for obtaining the three groups of experimental animals in this study.

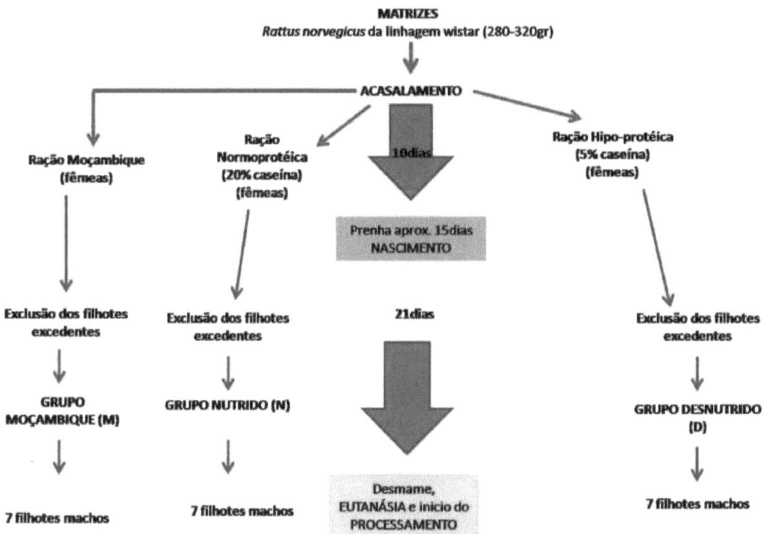

Source: Nhamposse, C. T. 2012

The animals used for mating were kindly provided by the Central Animal Facility of the Institute of Biomedical Sciences of the University of São Paulo (ICB-USP) and were kept in the animal facility of the Anatomy Department of the same Institute. All the animals used in this study (from mating to the end of the experimental period) were kept in air-conditioned rooms with a temperature of 22 ± 2° C and a 12-hour light/dark cycle. The breeding animals weighing between 280 and 320 grams were housed in polypropylene plastic cages measuring 41x34x16cm lined with autoclaved wood shavings, arranged on a vertical shelf during the adaptation (seven days), mating and suckling periods (Figure 6 A).

During the adaptation period, the boxes were changed daily. During this phase, the females that would form the nourished and malnourished groups were given unrestricted filtered water and normal rodent feed from the vivarium (pellets), and Mozambique feed for the animals that would form the experimental group.

Mating and pregnancy

In order to mate, the females were induced into the oestrus cycle through the "Whitten Effect", which is characterised by exposing the female to the odour caused by the male's urine, causing the female to release gonadotrophins and activating the ovarian cycle synchronously (WHITTEN, 1956). To do this, the female boxes were placed next to the male boxes for 2 days, after which the females were introduced into the male boxes (without the presence of males) for 2 days (without sanitising the boxes). During the same period, the males were housed in the female boxes. The males were then introduced into the boxes where they remained for ten days.

From the first day of mating, the standard animal house diet (pellets) was changed and the animals in the groups (Nourished = N, Malnourished = D) were given water and rodent feed (AIN-93G) *ad libitum* (REEVES; NIELSEN; FAHEY-JÚNIOR, 1993) and Mozambique feed (M), all prepared in a specialised laboratory (Rhoster Indústria e Comércio Ltda.) (Figure 6B).

After 10 days of mating, the females were separated from the males and placed in individual cages, according to diet, during gestation and lactation.

Obtaining experimental animals

During the experimental period, the animals were visited daily at 10am to replenish their water and feed and to clean their excreta (urine and faeces). Also during the experimental period, physical aspects were observed, such as locomotor activity, the appearance of wool (and later hair), the formation and appearance of the coat, the detachment of the ears, the opening of the eyes, as well as behavioural and physical aspects of the puppies.

Weaning took place at 21 days of age and the animals were euthanised for this experiment. Ten (10) pups were established as litters, and litters with less than 4 pups were discarded, litters whose mothers ate the pups and surplus litters with 8 pups were used as pilots for the various experimental techniques. The aim of reducing the litter size was to avoid the standard deviation in body weight and the high mortality rate that occurs in most malnourished litters due to protein deprivation, and also as a way of avoiding cannibalism which, when it occurs, requires the total exclusion of the litter and/or the female. The chicks used in the experiment were then identified, according to their diets, into groups N, D and

M, from which we chose male chicks used in this study. All groups consisted of 7 animals (5 animals for optical microscopy and 2 animals for ultrastructural studies using transmission microscopy). The pups' body weights were obtained at birth and this measurement was also taken at weaning. It should be noted that the experimental groups had already been formed since the adaptation period.

Sexing was carried out based on the ano-genital distance, which is much greater in the male than in the female (Figure 6C), and visualisation of the scrotum in the male chicks. The selection of these male chicks was also based on individual robustness and the white spot visible on the chicks' abdomens (Figure 6D).

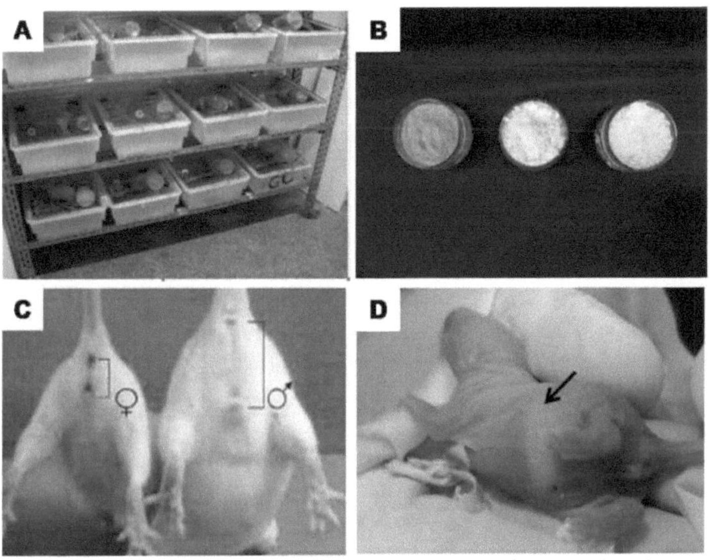

Figure 6- Conditioning, feeding, sexing and selection of the animals in the experiment

A shows the arrangement of the boxes on shelves in the animal house where the animals were kept to receive their diet. In B, the appearance of the three diets used in the experiment. From left to right: Mozambique, Normoprotelca and hypoprotein. In C and D, see, respectively, the difference between the anatomical anogenital distance, which is greater in males, and the white spot (arrow) on the chicks' abdomens.

This white spot, an indication of milk in the stomach, is an important discarding factor, since the weakest chicks are those that have not suckled (ANDRADE; PINTO; OLIVEIRA, 2002).

4.1.2 Composition of the feed used in the experimental groups

The AIN-93G feed[3] , formulated in a specialised laboratory for the growth, pregnancy and lactation phases of laboratory animals, was the feed of choice in this research, with the following basic composition: maize starch, dextrinised starch (90-94% tetrasaccharides), choline birtrate, casein, microfine cellulose, L-cystine, soya oil, mineral premix AIN-93G, vitamin premix AIN- 93G, SACAROSE, TERT-BUTYLHYDROCHINONE (REEVES et aL, 1993).

For the animals in the control group (or nourished - N), the percentage of casein was 20 per cent, while for the animals in the malnourished group (experimental group 1- D), this percentage was approximately 5 per cent.

With regard to the Mozambique ration given to the animals in experimental group 2 (M), its basic composition included Maize Flour, Roasted Peanut Flour (the vegetable protein component), Freeze-dried Cabbage, Mineral Premix AIN-93G and VITAMIN PREMIX AIN-93G.

Table 2 shows the components and their respective proportions in the three types of diets used in this study.

Table 2 - Composition of the standard (normoprotein), hypoprotein and Mozambique (M) diets

AIN-93G diet	normoprotein ration	hypoprotein feed	Mozambique ration
Humidity	8,4%	7,8%	4,75%
Crude protein	20% (casein)	5,2% (casein)	16,25%* (vegetable)
Ether extract	7,2%	7,0%	24,16%
Fibre matter	4,1%	5,0%	0,96%
Mineral matter	2,8%	2,8%	3,31%
Calcium	0,56%	0,5%	-
Phosphorus	0,26%	0,3%	-

* The protein in Mozambique's feed is vegetable protein.

4.1.3 Euthanasia and obtaining the gastrocnemius muscle

The animals in all groups N, D and M were euthanised using the chemical method of inhalation in a carbon dioxide (CO2) chamber located in the vivarium of the Institute of Biomedical Sciences (ICB III/USP). The choice of this method is based on the fact that CO2 is neither flammable nor explosive and is therefore safer, as well as being inexpensive. Its mechanism of action is fast and causes depression in the central nervous system, causing no apparent trauma to the animal (ANDRADE; PINTO; OLIVEIRA, 2002). In addition, this gas does not accumulate in the tissues and therefore leaves no residue.

[3] Rhoster Indústria e Comércio Ltda, SP, Brazil

The euthanasia method used for the transmission electron microscopy procedure was carried out according to the protocol of Ciena et al. (2010). The animals in all the experimental groups (N, D and M) were first anaesthetised intraperitoneally with 10% ketamine hydrochloride (50mg/kg) and 2% xylazine hydrochloride (10mg/kg) and kept in a prone position.

Dissection: obtaining the gastrocnemius muscle (MGc)

After making a lateral longitudinal incision in the skin of the pelvic limb, it was folded back medially in order to expose and remove the MGc. Each specimen was then fixed longitudinally by the common calcaneal tendon to a balsa wood base, with its belly surrounded by adhesive tape (durex), and covered in a small amount of special embedding medium for freezing biological tissues[4] , to keep it in the desired position (longitudinal). After adding neutral talc[5] to the surface of the specimens in order to avoid possible artefacts resulting from the low temperature, they were completely covered in the embedding medium and immediately placed in a stainless steel beaker containing isopentane, which was immersed in liquid nitrogen until it reached a temperature of approximately -150°C (Figure 7). The samples were stored in a freezer at -80°C (Thermo Forma: -86C ULT Freezer) in the multi-user laboratory of the Anatomy department at ICB/USP. The material thus obtained was subsequently processed in the Laboratory of Functional Anatomy Applied to Clinical and Surgical Practice (LAFFAC) of the Anatomy Department of ICB III/USP.

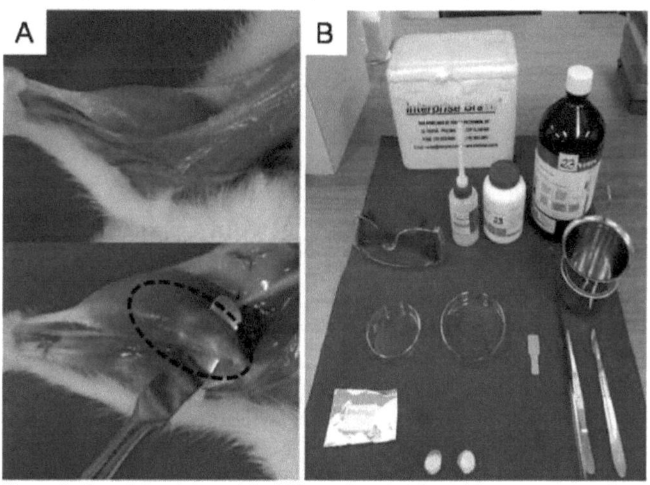

Figure 7- Dorsolateral view of the MGc and material used for cryopreservation of the
Caption: Dissection of the gastrocnemius muscle (A) and material used to cryopreserve the samples (B).

[4] KILLIK - EasyPath®, São Paulo, SP, Brazil
[5] Tragacanth® (Sigma)

4.1.4 Routine histology procedures

After being removed from the freezer, the specimens were first kept in the cold room of a cryostat (Leica CM 1850) at -25°C, where they remained for approximately 30 minutes, a period sufficient for them to adapt to the new temperature. Cross-sections along the long axis of the MGc were obtained serially and 10pm thick, according to the stereological criteria established by Howard and Reed (2005), and placed on slides stored at room temperature until they were submitted to the different histological and histochemical staining methods described below.

4.1.4.1 Haematoxylin-Eosin (HE)

This method, recommended by Behmer; Tolosa and Freitas-Neto (1975), was used to highlight the tissue cytoarchitecture of the MGc. The sections were first washed for 2 minutes in distilled water to remove excess powder from the inclusion medium and then immersed for 4 minutes in haematoxylin solution. After removing the excess dye with distilled water for a few minutes, the sections were submerged in eosin solution for 2 minutes, after which they were bathed again in distilled water and dehydrated in 95% ethanol and absolute ethanol for two minutes each. After being diaphanised in xylene (3 passes of 2 minutes each), the sections were mounted with Entellan[6] between slide and coverslip and analysed and photographed on an image analyser system[7] belonging to LAFACC.

4.1.4.2 . Picrosirius

The collagen fibres were visualised in sections stained using the Picro-Sirius method and analysed under polarised light, as recommended by Junqueira, Bignolas and Brentani (1979). Initially, in order to remove excess talc, the sections were washed in distilled water for minutes and immersed in the Picro-Sirius solution for 20 minutes. After washing in distilled water for 5 minutes, they were dehydrated, diaphanised, mounted and analysed as described in the previous section.

4.1.4.3 Nicotinamide Adenine Dinucleotide Tetrazolium Reductase (nadh-tr)

The different types of muscle fibres were characterised according to metabolic and functional parameters using the NADH-tr histochemical reaction (Sigma, N8129), according to the protocol established by Barnárd et al. (1971) and Silva (1999). This staining method uses the oxide-reduction reactions of the Krebs cycle. Through enzymatic action, Nitro Blue Tetrazolium (NBT), a yellow salt, reduces, becomes insoluble and precipitates, giving rise to a blue-violet compound, formazan. The enzyme responsible for this reaction is present in

---

[6] Entellan ® (Merck)

[7] Zeiss Microimaging Cari, model Stemi® SV6 - LAFACC - ICB/USP

the internal fibril regions, mitochondria and sarcoplasmic reticulum of muscle cells. The intensity of this reaction reflects the oxidative capacity in the cell's metabolism, allowing muscle fibres to be distinguished into types I, HA and IIB or O, GO and G, respectively. Therefore, oxidative fibres (O or type I), because they contain these structures and organelles in greater concentration, have intense oxidative activity and are coloured more intensely in dark blue (SILVA, 1999); however, type IIB (G) fibres, which depend more on anaerobic glycolysis, appear in a lighter shade of blue and type HA (GO) fibres, with glycolytic and oxidative metabolism, show intermediate colouring. In addition to their colouring, O fibres have a small diameter, GO fibres an intermediate diameter and G fibres a large diameter.

The slides were then incubated for 40 minutes at 37°C in a solution containing Tris buffer (0.2M; pH 7.4), 5mg of NBT (Sigma, N6876) and 12.5mg of NADH-tr. After incubation, the slides were immersed in a sequence of acetone (60%, 90% and 60%), alcohol (from 80% to absolute) and diaphanised in xylene (3X2 minutes), then mounted with Entellan (Merck) between slide and coverslip and analysed using the computerised image acquisition system .[8]

4.1.4.4 Transmission electron microscopy (TEM)

To prepare the samples for MET, after euthanising the animals from all the experimental groups (N, D and M), they underwent thoracotomy followed by cardiac ventricular perfusion (left ventricle) using a modified Karnovsky fixative solution (2% paraformaldehyde and 2.5% glutaraldehyde in cacodylate buffer solution, 0.1 M; pH 7.4).

After perfusion, the MGc were dissected and fragments of approximately 2mm were removed and immersed in a fixative solution of 2.5% glutaraldehyde in cacodylate buffer, where they remained for 2 hours at 4°C.

The fragments were then washed three times in cacodylate buffer solution and post-fixed in 2% osmium tetroxide (OSO4) in cacodylate buffer solution for 2 hours at 4°C. After washing in cacodylate buffer, the specimens were treated with uranyl acetate solution (0.5%) where they remained protected from light for 12 hours at room temperature. After this period, the water in the fragments was replaced by immersing them in a solution of propylene oxide (miscible in both water and resin), and they were then dehydrated in an increasing series of ethanol (70% to absolute) and pure propylene oxide.

[8] Power shot A640, Canon, China-LAFACC/USP

After dehydration, the specimens remained for 4 hours in a mixture of resin (Low Viscosity Embebbing Medium Spurr's Kit, Electron Microscopy Sciences, USA) and propylene oxide in a ratio of 1:1 and kept under agitation[9] at room temperature, for gradual infiltration of the resin into the material. The mixture was then replaced with pure resin, where the specimens remained for 12 to 18 hours under agitation at room temperature. After this period, the resin was again replaced with pure resin, and the specimens were kept for 1 hour in an oven at 37°C, with the bottle open to allow the liquids to evaporate completely from the material.

Inclusion was carried out by placing the material in rectangular moulds filled with pure resin and kept for five days in an oven at 60°C to allow the block to polymerise. The blocks thus obtained were cut into semi-thin sections (400nm thick) using an ultramicrotome (Leica Ultracut UCT, Germany) and then stained with 1% toluidine blue to check the integrity of the material under light microscopy. Subsequently, once the area of interest had been selected, ultra-thin sections (40nm thick) were made on ultramicrotome, which were collected on suitable 200 mesh copper screens (Sigma-Aldrich, USA) and counterstained with uranyl acetate solution (4%) for 3 minutes, washed in distilled water, contrasted with aqueous lead citrate solution (0.4%) for 3 minutes and washed again in distilled water (REYNOLDS, 1963).

The screens were analysed and photographed using a transmission electron microscope (FEI Morgagni 268 (D) Netherlands) at the Anatomy Department of the Faculty of Veterinary Medicine and Zootechny of the University of São Paulo (VCI - Anatomy/FMVZ-USP).

4.1.4.5 Morphometric evaluation of the gastrocnemius muscle

Cross-sectional area of muscle fibres

The sections stained using the NADH-tr histochemical method (item 4.1.1.3) were used to determine the area of the muscle fibre profile (ASTf) of types I (O), IIA (GO) and IIB (G). For each animal in each group (N, D and M), the areas of 10 fibres of types O, GO and G were obtained, making a total of 50 fibres of each type for each group. Images were acquired using a stereoscopic magnifier [10][11] , with a digital camera attached[11]. The images were processed using computerised imaging equipment[12] (Figure 8).

---

[9] Analíticaó - Nova Analítica, SP, Brazil
[10] Zeiss Microimaging Cari, model Stemi® SV6 - LAFACC - ICB/USP
[11] Power shot A640, Canon, China - LAFACC - ICB/USP
[12] Axiovision Rei. 4.6, Gõttingen, Germany - LAFACC - ICB/USP

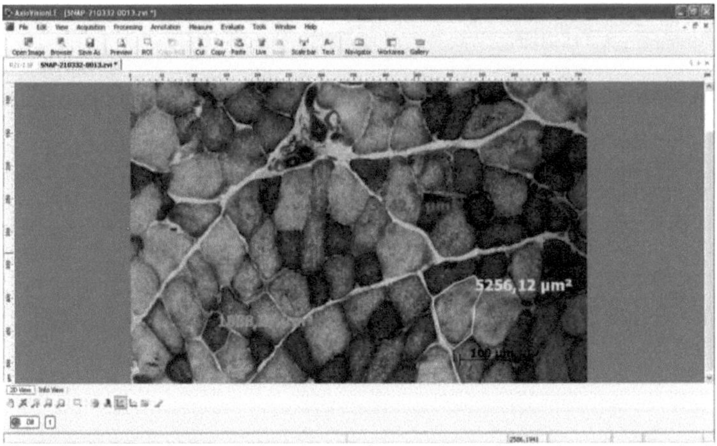

Figura 8- Cutaway image of rat gastrocnemius muscle obtained using a computerised analysis system. Analysis of fibres stained using the NADH-tr method, showing the cross-sectional area of type I (O) fibres in green, IIA (GO) fibres in red and IIB(G) fibres in yellow.

4.1.4.6 Muscle fibre density

To determine the density (in percentage) of muscle fibre types, a 120-point test system was used, which was placed over 3 images of each animal in each group, obtained from sections stained by the NADH-tr histochemical method, using the computerised imaging system described in the previous section (Figure 9).

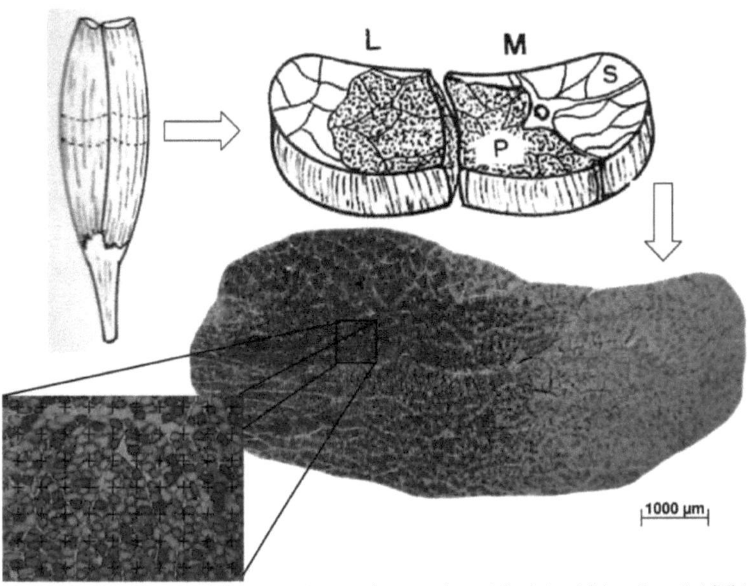

Figura 9- diagram of the rat gastrocnemius muscle and the lateral (L) and medial (M) heads

Adapted from Alves, 2010. The images show a section stained using the NADH-tr method, from which an image has been highlighted with the Cross Test System superimposed on it.

The results of the quantitative assessment were expressed in tables showing the mean ± standard deviation and then represented in graphs. All results were statistically analysed using the Anova and Tukey statistical tests.

CHAPTER 5

RESULTS

5.1 QUANTITATIVE ANALYSIS

5.1.1 Animal weight

5.1.1.1 Birth weight

Table 3 shows the average birth weights of the animals; there was a significant difference between the groups, $p<0.001$. The animals in the Mozambique group had a higher birth weight than the nourished and malnourished ones; the nourished ones had a higher birth weight than the malnourished ones (Figure 10).

Table 3- Birth weight (g) of the three groups

	Nourished	Malnourished	Mozambique
Average	6,61	5,50	7,38
DP	0,59	0,23	0,38

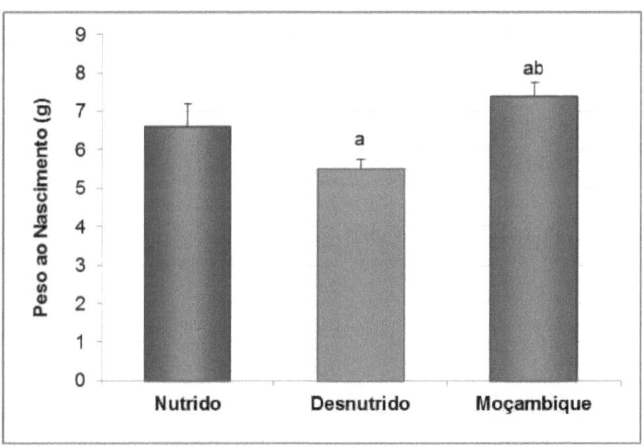

Figure 10- Graph of the average birth weights of the animals in the three study groups
The animals in group M weighed more than those in groups N and D: vs. Nourished; b: vs. Malnourished, $p<0.05$.

5.1.1.2 Weaning weight (21 days)

Table 4 shows that there was a significant difference between the average weights of the three groups ($p<0.001$) at 21 days. The nourished animals had a higher weight gain and had the highest weight, followed by those from Mozambique and finally the malnourished ones (Figure 11).

Table 4- Weight (g) of the three groups at 21 days

	Nourished	Malnourished	Mozambique
Average	59,08	23,15	35,49
DP	4,25	4,04	3,01

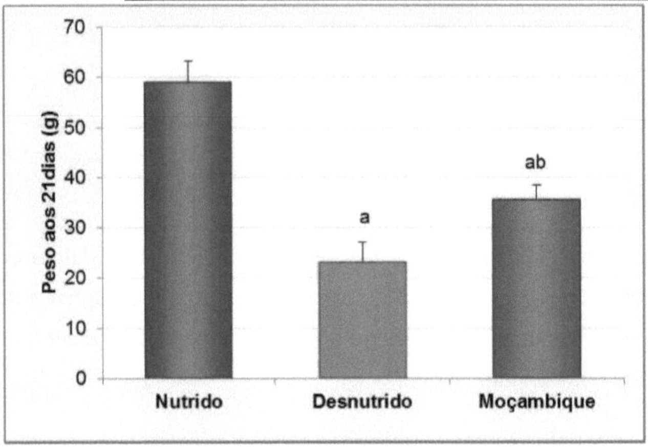

Figure 11- Graph of the average weaning weights (21 days) of the animals in the three groups
It can be seen that the animals in group N gained more weight than those in groups M and D; a: vs. Nourished; b: vs. Malnourished, p<0.05.

5.1.1.3 Length at birth

Table 5 shows that there was a significant difference (p<0.001) in the average length at birth; in fact, the length of the Mozambique animals was close to that of the malnourished animals (there was homogeneity in the lengths of these two groups), but both were shorter than the length of the nourished animals (Figure 12).

Table 5- Average length (cm) of the three groups at birth

	Nourished	Malnourished	Mozambique
Average	5,96	5,24	5,37
DP	0,11	0,04	0,10

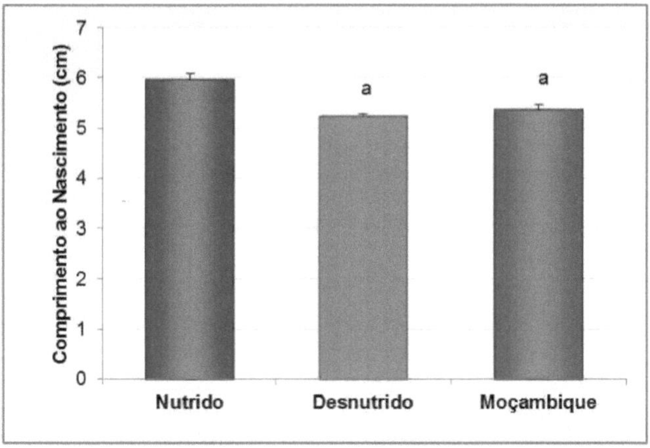

Figure 12 - Graph of the average lengths at birth of the animals in the three groups.
Note that there is no significant difference between the length of the animals in groups D and M; a: vs. Nourished; b: vs. Malnourished, p<0.05

5.1.1.4 Length at weaning

Table 6 shows the average lengths of the animals at 21 days; there was a significant difference between the groups (p<0.001); the nourished animals had the greatest length, followed by the Mozambique animals and finally the malnourished ones (Figure 13).

Table 6: Length (cm) of the three groups at 21 days

	Nourished	Malnourished	Mozambique
Average	12,80	9,30	10,28
DP	0,56	0,53	0,42

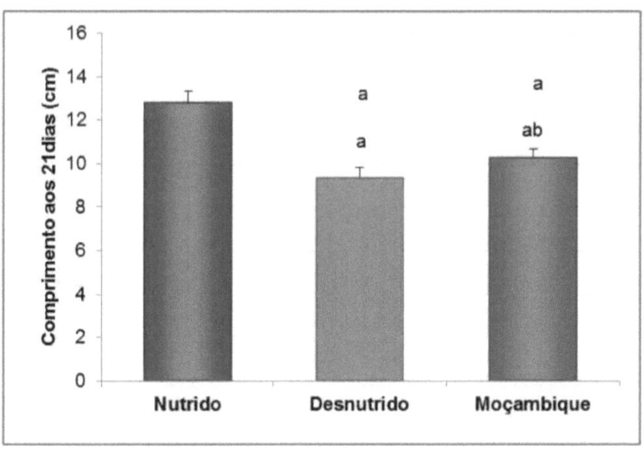

Figure 13: graph of the averages for length at weaning. Group N had the longest length at weaning, followed by group M and finally group D; a: vs. Nourished; b: vs. Malnourished, p<0.05

5.1.2 Muscle Fibre Density (MFD)

The density of muscle fibre types is expressed as a percentage in figure 14.

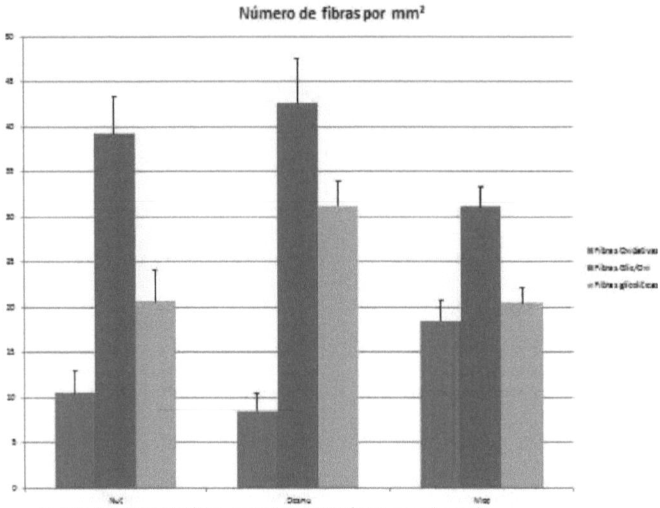

Figure 14- Muscle fibre density graph (in percentages)

The percentage of oxidative fibres was higher in group M compared to groups N and D, which did not differ ($p<0.05$). There was no difference between the groups in the percentage of GO fibres ($p<0.116$). The animals in group D had a higher percentage of G fibres than groups N and M, which were the same ($p<0.05$).

Thus, initially there was a statistical difference in relation to nutritional status ($p<0.05$). With regard to glycolytic/oxidative fibres, the percentage of

of these compared to the others (oxidative and glycolytic) was higher in all groups, with no predominance between them ($p<0.116$).

The percentage of glycolytic fibres was higher in the D animals than in the N and M groups, while the percentage of oxidative fibres was higher in the M group than in the N and D groups.

5.1.3 Cross-sectional area of muscle fibres (ASTf)

The results obtained for the ASTf of the glycolytic, glycolytic-oxidative and oxidative fibres of the different groups, respectively, are shown in Figures 15, 16 and 17.

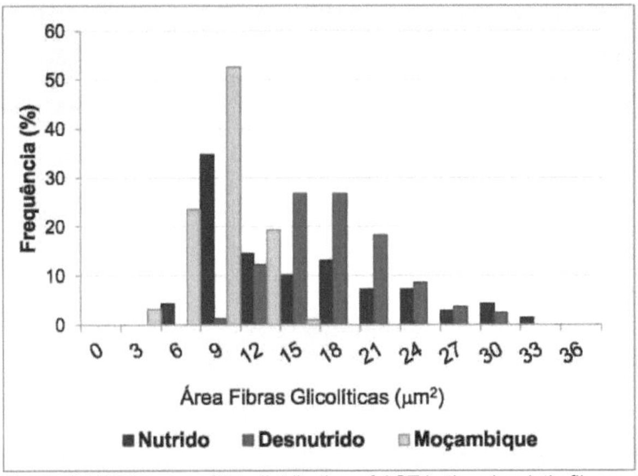

Figure 15- Histogram of the distribution of AST in the glycolytic fibres of rat MGc

Figure 15 shows a tendency for animals in groups N and M to have lower ASTf of glycolytic fibres compared to animals in group D.

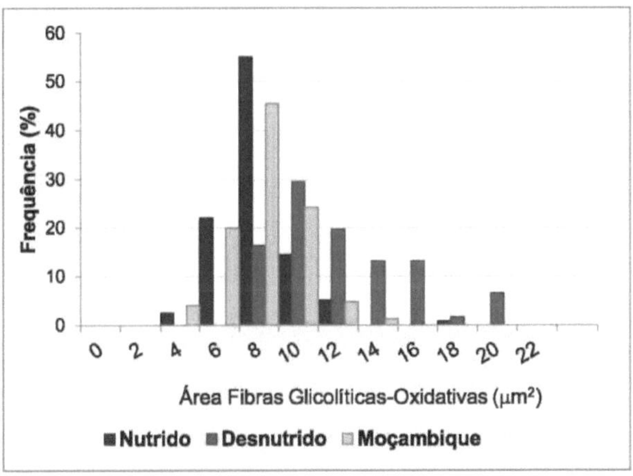

Figura 16- Histogram of AST distribution of Glycolytic-Oxidative fibres of rat MGc

Figure 16 shows that the highest concentration of the ASTf of the glycolytic/oxidative fibres of the animals in groups N and M is between 5 and 11 μm^2, while the ASTf of these fibres in the animals in group D is distributed between 8 and 16 μm^2.

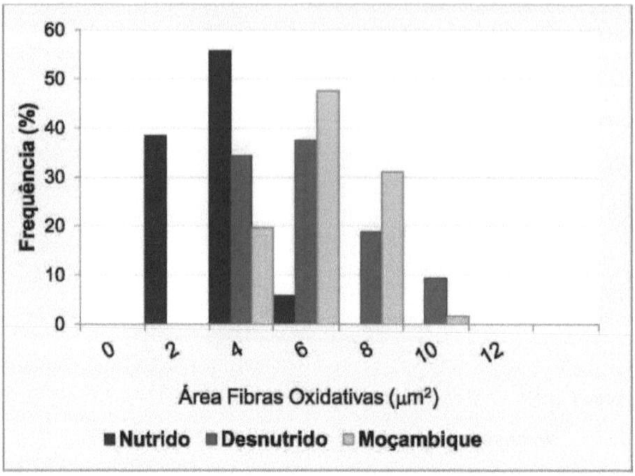

Figura 17- Histogram of AST distribution in the oxidative fibres of rat MGc

Figura 18-

Figure 17 clearly shows the concentration of ASTf in the oxidative fibres of group N at lower levels than in groups D and M, which did not differ from each other.

Table 7 shows the average ASTf (± SD) for the different types of fibres in the three study groups.

Table 7: Cross-sectional area of the muscle fibres (pm^2) of the rat MGc

Type of fibre	Nourished Group (N)	Group Malnourished (D)	Group Group Mozambique (M)
Glycolytic (G)	16,3±6,0	19,4±2,1	10,3±0,8
Glycolytic/Oxidative (GO)	8,9±0,9	**13,8±2,1**	9,2±0,7
Oxidative (0)	4,4±0,5	7,0±0,9	7,5±0,7

Analysis of these data shows that, as far as glycolytic fibres are concerned, group M had the lowest ASTf when compared to groups N and D. There was no statistically significant difference between groups M and D when assessing oxidative and glycolytic-oxidative fibres (although these were slightly higher in M than in D). When comparing the data for the Glycolytic fibres with the control group (N), it was noted that the ASTf of the animals in group D was higher, while there was no difference between the ASTf of the control and group M. For the oxidative fibres, the lowest ASTf was detected in the animals in group N, while for those in groups D and M, no significant differences were detected between the two. However, when analysing the AST of the fibres of the Mozambique animals in isolation, we could see that they did not differ significantly from each other compared to the animals in groups N and D, which showed large differences in their ASTf.

5.2 QUALITATIVE ANALYSIS

5.2.1 Histology

HE staining revealed structural differences between the muscles of the animals in the three groups (N, D and M) (Figure 18A, Be C). Thus, the animals in group N had the majority of their muscle fibres with a polygonal shape and a high density of intensely stained nuclei located at the periphery of the cell. A well-defined and relatively thick endomysium was observed surrounding each fibre in isolation, giving the section a compacted appearance.

In the animals in group M, the muscle cells were compacted in their fascicles, similarly to what was observed in group N. However, in terms of their outline, many of them were rounded. The nuclei, located peripherally, were well-defined, but their contour was clearly altered, i.e. they were dilated in comparison to those in group N. Also, when compared to the control group, these nuclei had an apparently lower density. The endomysium, although organised, was thinner than that present in the nourished animals

The animals in group D clearly lost the compacted appearance of the cut seen in group N. The cells, which were more widely spaced, not only had different shapes but also a total discrepancy in size, ranging from small rounded cells to large oval or irregular cells. Their nuclei, although still located peripherally, were slightly more elongated and less coloured than those of the animals in group N. In relation to the connective tissue, there was a much thinner endomysium, which was difficult to detect in some cells that showed a loss of cell contours.

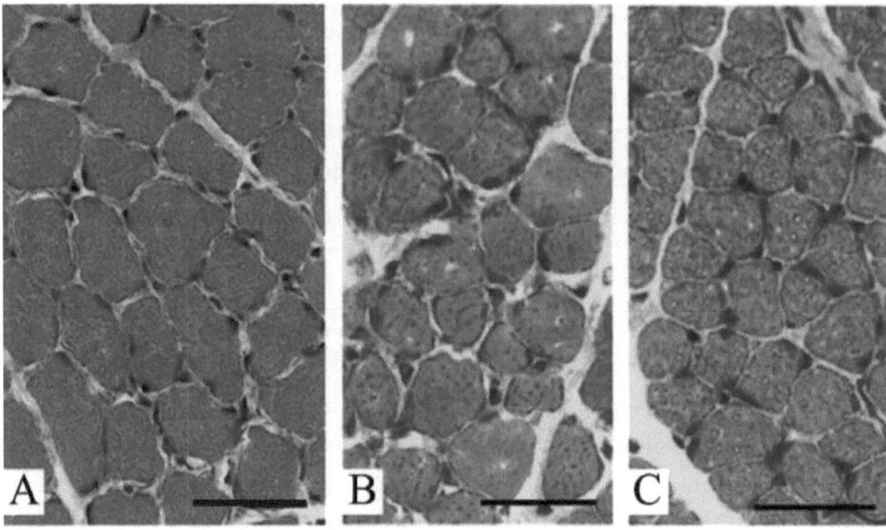

Figura 18- Photomicrograph of rat gastrocnemius muscle

Note in A - animals from the nourished group (N), the muscle fibres are polygonal in appearance and densely arranged, with a regular and relatively thick endomysium. Note the flattened and densely packed nuclei. B- Animals from the malnourished group (D), showing cells of different sizes and shapes, spaced apart and with irregularly distributed dilated nuclei, as well as disorganised endomysium. C- The animals in the Mozambique group (M) are characterised by compacted muscle fibres, but with rounded contours. The peripheral nuclei are dilated and less numerous, and the endomysium is extremely thin when compared to the control group (N). (HE colouring. Calibration bar: 200 μm).

When the sections from the different groups (N, D, M) subjected to Picro-sirius staining were compared and analysed under polarised light (Figure 19 A, B and C), it was found that, taking the perimysium as a parameter, this envelope was made up almost exclusively of type I collagen fibres (shown in red, orange and yellow). In the groups of malnourished animals (D) and those subjected to the Mozambique diet (M), the difference was clear when compared to the control group (N), since the perimysium of the muscle of these animals was clearly dominated by type III collagen fibres (green).

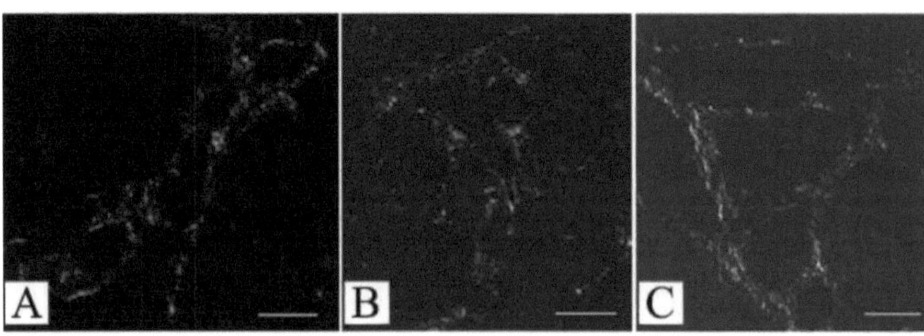

Figure 19: Sections of the gastrocnemius muscle of animals from the Nourished (A), Malnourished (B) and Mozambique (C) groups showing the constitution of the perimysium.
Note in A, the predominance of type I collagen fibres (red, orange and yellow), and in B and C, the predominance of type III collagen fibres (green). Bar: 20pm

5.2.2 Ultrastructural analysis (MET)

In this assessment, the muscles of the animals in group N had a well-defined morphology; thus, the myofibrils and Z-lines were aligned; the intramyofibrillar spaces were preserved and contained sarcoplasmic organelles, the mitochondria. The ratio between bands A and I was approximately 3:1 (Figure 20 A, B).

The animals in group D had broken myofibrils or misaligned sarcomere Z-lines. There was an increase in band I (Figure C, D, E). The animals in group M also had ruptured myofibrils and misaligned Z-lines, as well as increased intomyofibrillar space with the presence of larger mitochondria (Figure 20 F, G). Also in the M group, with regard to the A-band/I-band ratio, there was an increase in the I-band, reaching a ratio of 2:1, similar to the malnourished animals.

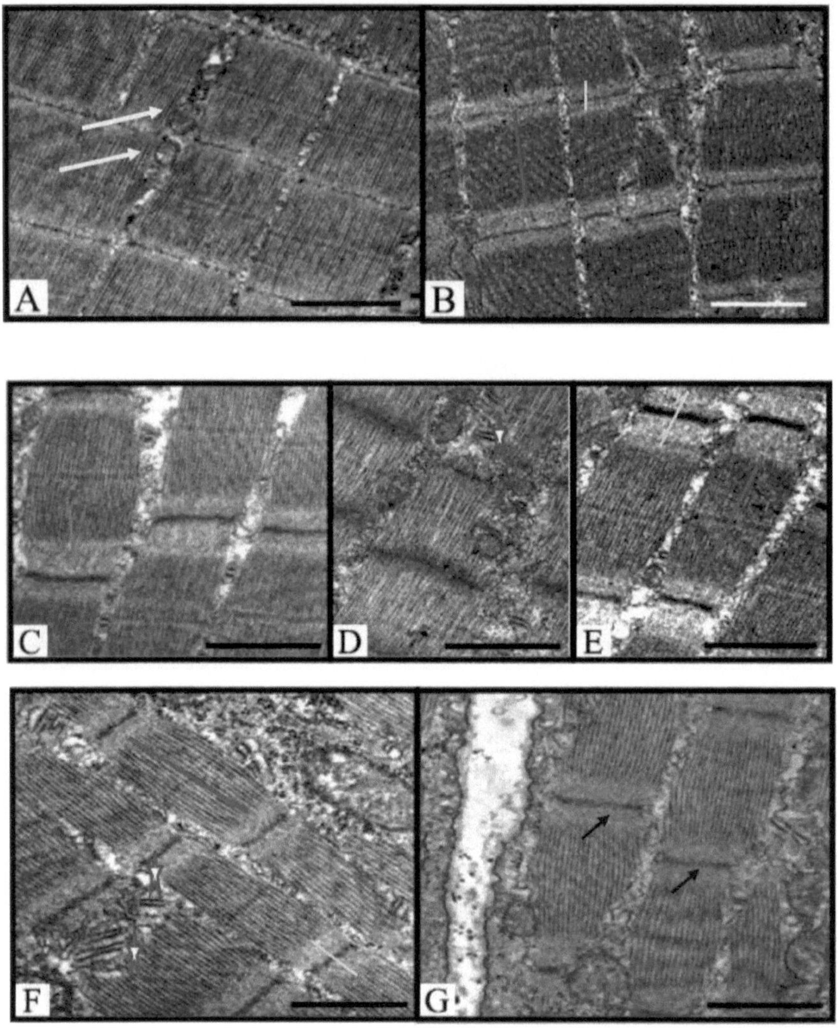

Figure 20 - Electron micrograph of rat gastrocnemius muscle

A and B- Nourished animals, showing myofibrils and well-aligned Z-lines; the intermyofibrillar spaces were preserved and contained some sarcoplasmic organelles such as mitochondria (arrows). The ratio between bands A and I was approximately 3:1. C, D, E- Malnourished animals showing misalignment of the sarcomere Z-lines (green arrows) with rupture of the myofibrils (*). Note the increase in Band I, such that the ratio between Band A and I is close to 2:1. F, G_ Animals from Mozambique, showing misalignment of the Z lines (short arrow), and rupture of the myofibrils. intermiofibrillar space with the presence of some mitochondria of different sizes; band A/band I ratio similar to that of group D at 2:1 (long arrow). Band A (red), Band I (yellow). Bar: 1µm.

CHAPTER 6

DISCUSSION

6.1 GENERAL CONSIDERATIONS

Children form a group of the population that is particularly susceptible to malnutrition, as they need a greater quantity of calories and nutrients for their growth and development. When the nutrients ingested are insufficient for the body to function, it uses the energy stored in glucose in the tissues and blood, glycogen in the liver and muscles and, lastly, liver and muscle proteins. In order to reverse malnutrition, in addition to the need to ingest certain vitamin complexes, it is essential to adopt a balanced diet in order to maintain the essential amount of water in the body and to combat weight loss. Although CD is the biggest cause of infant mortality in Mozambique, recent studies show that there is almost no information on the nutritional requirements for preparing a balanced diet using local cultures for children with CD (BANDEIRA et al, 2001; ROSE; TSCHIRLEY, 2003; PRISTA et al.,2003HAMPSHIRE et aL, 2004; ROSENTHAL et aL, 2004; LOW et aL, 2007), a fact that justifies this study, emphasising the aim of testing whether the introduction of micronutrients into the ration prepared with crops produced in Mozambique will increase the effectiveness of that diet by adopting a cheap and nutritionally effective method of a balanced diet.

Several studies have shown that malnutrition affects systems such as the CNS (WINICK, 1969; HEDLEY-WHITE; MEUSER, 1971; BEDI; WARREN, 1983; CORDERO et aL, 1986), bone tissue (SHRADER; ZEMAN, 1973) and joints (SPANHEIMER et aL, 1991).

In the present study, muscle tissue was chosen because its development is determined by the individual's nutritional status (HAJ et al, 1986) and is therefore directly affected by CD, as reported by Waterlow (1955; 1956) who measured DNA in the muscle and liver of patients with kwashiorkor and found that the rate of morphological changes was higher in the muscle than in the liver. The authors concluded that in kwashiorkor the major site of tissue protein depletion is the muscle, plus the fact that this organ makes up half of the body mass (STANDRING et aL, 2008).

According to Rose and Rothstein (1982), muscle tissue has the ability to adapt easily to changes that modify its functionality, which is why understanding the morphological changes caused by malnutrition in this tissue can help in its rehabilitation.

Few morphological studies on muscles in food deprivation have been carried out to date (DASTUR et aL, 1982; OLFORDS et aL, 1983). Among the various alterations reported in skeletal muscles in cases of malnutrition, the following stand out: Hedley-White and Meuser (1971), Hedley-White and Meuser (1973), Sima (1974), Bedi and Warren (1983), Conboy et al. (1987) and Cornblath and Brown (1988), which dealt with the restriction of food supply, Roy et al. (1972), Oldfords and Persson (1982), Conradi et al. (1985,1989) and Rana et al. (1991) who used a low-protein diet, Brandão (1998) and Castelucci (1999) who used a high-protein diet, or Winick and Noble (1966) and Gopinath et al. (1983) who used a concomitant restriction of food supply and litter expansion. These authors were unanimous in reporting muscle atrophy and a reduction in the diameter of the muscle fibres, preferably type II, as "typical" changes in these cases (VAN BOGAERT et al., 1962; MONTGOMERY, 1962; HARRIMAN, 1965; SACHDEV et al., 1971; DASTUR et al., 1975; HANSEN-SMITH et al., 1979; ESSEN et al., 1981; LINDBOE et al., 1982; OLDFORDS; SOURANDER, 1986). In addition, Simão (2004), Mathiensen (2004) and Pacheco (2006) reported an increase in fibre section density concomitantly with a decrease in fibre profile area in cases of protein deprivation.

Although Montgomery (1962), Enesco and Puddy (1964) and Oliveira et al. (1999) have determined that the effects of early malnutrition on muscle development may be permanent, supporting what has been previously reported in previous studies (MONTGOMERY, 1962; ROY et al., 1972; GOPINATH et al., 1983), in which it was also shown that in states of malnutrition there is an irreversible reduction in the number of muscle fibres, Oldfords and Sourander (1986) and Jahoor et al. (2008) concluded that there was no irreversible damage to the muscles of protein-deprived rats, supporting the previous findings of authors such as Sima (1974), who observed that the reduction in fibre diameter was partially reversed in renourished animals, although there was still a reduction in fibre growth.

Information on the long-term effects of child malnutrition is still inconsistent, as the length of time the individual has been exposed to malnutrition must be taken into account. Oldfords and Sourander (1986) and Jahoor et al. (2008) show in their studies that when malnourished children are treated in time, there is complete recovery of organs such as the liver, muscles and immune system. Furthermore, Nascimento (1990) points out that in some cases food deprivation occurs so "drastically" that there is no time for morphofunctional reflection. However, Martins (2008) and Azzari (2010) point out that in some cases, recovery is almost impossible, as there is impaired absorption of nutrients by the intestine and varying degrees of mental impairment, symptoms which are associated with the length of time the child has been malnourished, the severity of the illness and the age at which the illness began.

In this study, in order to get as close as possible to the "real" dietary situation of populations in rural and peri-urban areas of Mozambique, Wistar rats were used as an experimental model and fed a ration based on peanuts, maize and cabbage, with the aim of verifying the morphological changes that occurred in the MGc of these animals at weaning (21 days) and comparing these parameters with rats of the same age, but which in the meantime were fed a normoprotein (nourished) and hypoprotein (malnourished) diet formulated from AIN-93G food prepared in a specialised laboratory (Rhoster Comércio e Indústria Ltda, SP, Brazil). Two parameters are used to measure malnutrition in Mozambique: height for age and weight for age (MARTINS, 2008; RELATÓRIO, 2011).

Based on the information in the literature, the work by Nascimento (1990) was the closest to the present study. In fact, this author subjected albino rats to a state of protein deficiency (6.7% protein) during the gestational period and also the pups, which were sacrificed at 15 and 30 days respectively.

The findings of Brooke and Kaiser (1970) and Brooke, Williamson and Kaiser (1971) served as a basis for extrapolating the findings of the present study to humans, since these authors proved that the type of fibres found in the rat's gastrocnemius muscle is similar to the types of fibres in the same muscle in humans.

6.2 QUANTITATIVE ANALYSIS

In order to assess the extent to which the Mozambique diet was able to alter the animals' **body mass**, and whether or not it was associated with malnutrition, it was considered important in this study to measure weight and body length during the study period. According to Martins (2008) and the Report (2011), weight for age and height for age (which in children under 2 years of age and with a length < 85 cm is considered length for age) are considered to be the indicators of CD in Mozambique.

In the present study, the **length of the animals** in group M was the same as that of the malnourished animals and their body mass, although higher at birth, decreased over time, becoming lower than that of the nourished animals at weaning. These observations are similar to those made by Zeman (1967) when he reported that the body weight and survival rate of pups from females fed a hypoprotein diet (6% casein) were lower than those of pups from females fed a diet containing 24% casein (hyperprotein) during gestation. Olfords and Sourander (1986), who fed rats with a low percentage of protein (malnourished), reported a reduction in weight, reaching 53% less than controls of the same age, confirming our findings, but we also note that the findings of the present study support more recent evidence (NASCIMENTO, 1990) where albino rats were fed malnourished feed from mating, gestation and pups. This author found a weight reduction of up to 50% less than controls at the same age.

Based on the findings of the aforementioned studies, compared to the results presented here, we can explore the possibility that the decrease in weight during growth observed in the animals from Mozambique may have been due to the lack of protein in the feed, which led them to a state similar to that of the animals in group D (malnourished), as the two groups of animals showed the same behaviour in terms of body mass and length (although the weight of the animals from Mozambique was relatively higher than that of the malnourished animals). Similarly, Oliveira (2006) proposed the possibility that once the animals are malnourished, they no longer have any tissues to degrade, which could result in a big difference in their body mass.

In addition to this fact, there is evidence that peanuts (which were the basic component of group M's diet), due to their high energy content, increase the feeling of satiety, thus leading to a quick end to the meal and a consequent reduction in intake (KIRKMEYER; MATTES, 2000; ALPER; MATTES, 2002), thus proving that the calories from peanuts are not fully absorbed, leading to an increase in lipids and polyunsaturated fatty acids, promoting a lower than expected weight gain (SALLES et al., 2005; COELHO et al., 2006).

As for **muscle fibre density,** it was observed that the corresponding fibres differed in size and reaction to protein between the groups. With regard to glycolytic fibres, the animals in group M had the same proportion as N, but a lower proportion than the animals in group D; with regard to glycolytic/oxidative fibres, it was observed that the animals in group M had the lowest percentage compared to groups N and D. However, with regard to oxidative fibres, the animals in group M had the highest percentage compared to those in groups N and D. These results corroborate the findings of Oldfords and Sourander (1986), who described type II (GO) and IIB (G) fibres as being more susceptible to protein deprivation than type I (O) fibres, with a decrease in type II (GO) fibres compared to type IIB (G) fibres. What's more, the authors report that type II fibres (GO) are more influenced by nutritional supplementation, supporting our results where it was found that Mozambique's nutrition mainly affected intermediate fibres (GO), i.e. those with the lowest capacity to convert into G or O. These observations had already been reported in previous studies by

Montgomery (1962), Roy et al. (1972) and Gopinath et al. (1983), where it was determined that malnutrition caused an irreversible decrease in the number of muscle fibres, especially those of type II (GO). In a more recent study, Nascimento et al. (1990) found that in malnourished 15-day-old pups born to albino rats that had been subjected to protein deprivation during gestation and lactation, type II (GO) fibres were strongly compromised and observed dominant type I (O) fibres and smaller type II (GO) fibres and less than half of the total fibres.

Venema and Overweg (1974) studied 10 parameters for measuring the size and shape of cross-sections of human muscle fibres processed by fresh histochemical methods and concluded that all of them had imperfections, mainly due to the obliquity of the cut section in the muscle fibre. However, they admitted that the measurement of the area (AST) is an isolated parameter, but adequate for sizing the muscle fibre, a method that was used here.

Thus, when comparing the **ASTf** of glycolytic, glycolytic/oxidative and oxidative fibres, it was found that the animals in group M had a smaller diameter of G fibres compared to groups N and D. Previous studies by Timson and Dudenhoeffer (1985) showed similar results to ours where young rats that were subjected to severe protein restriction showed a decrease in muscle growth rate as well as in the size and number of muscle fibres. Nascimento (1990), who carried out a similar study to the present one, described a predominance of smaller calibre type II fibres (GO) than in the controls, but the reduction in GO fibres was more noticeable, as these were strongly compromised, as was the case in the present study, and Oliveira (2006) who found that the ASTf of type I, II, HA and IIB fibres was lower in the malnourished animals. In addition, Oliveira et al. (1999) found atrophy of type II fibres (GO) and clustering of type I fibres (O) in adult patients with malnutrition. These authors explain that the selective atrophy of type II fibres occurs in muscle myogenesis, because type I (slow) fibres are composed of a large proportion of muscle fibres from the primary myotubes, which are resistant to environmental factors, while type II (fast) fibres have a large proportion of muscle fibres from the secondary myotubes, which are therefore very susceptible to nutritional problems and neutrophic factors. Similar findings have been described in humans and rats. These reports lead us to hypothesise that in the present study the proportion of G fibres in group M was compromised during myogenesis, as they were smaller than in groups D and N, and there was also a decrease in the proportion of GO fibres, which were the main ones affected, as can be seen in table 5. This shows that nutrition is a factor that influences the cross-sectional area of the muscle studied.

6.3 QUALITATIVE ASSESSMENT

The normal compacted appearance of cells with intensely stained peripheral nuclei observed in the animals in group N was quite different from the animals in groups M and D. In the animals from group M, the muscle fibres showed a rounded outline, different shades of granular or acidophilic staining, more elongated and weakly stained nuclei, some of which had lost their cell outlines. Authors such as Bablet and Normet (1937) and Lindboe and Presthus (1985) associated the rounded appearance of the fibres and the increased intercellular space observed in the M group animals with muscle atrophy. A marked reduction in fibre thickness and a slight increase in the interfascicular space with a reduction in fibre calibre was observed by Wainio et al. (1959) and Walsh et al. (1971) and Burke and Tsaris (1974) in MGc in malnourished rats. Montgomery (1962) reported that in the sartorius muscle of Jamaican children who died of CPS at 8 months and 16 months of age, there was a reduction in muscle calibre and an increase in interfascicular connective tissue; these fibres reached a size comparable to that of foetuses between 31ª and 36ª weeks. On the other hand, Nascimento (1990) also found a reduction in fibre thickness and a slight increase in interfascicular space in albino rats at 15 and 30 days after weaning.

The results found by these authors confirm those described here, where a variation in the calibre of the muscle fibres was observed, with a reduction in these fibres and muscle atrophy in the animals in the M and D groups. The reduction in fibre thickness and muscle atrophy were previously reported by Vincent and Radermecker ([19-?]), Montgomery (1962), Enesco and Puddy (1964), Roy et aL (1972), Maki and Campos (1975), Gopinath et aL 1983, Haj et aL (1986), Oldfords and Sourander (1986), Nascimento et aL (1990) and Oliveira et aL (1999), reaffirming the results we found; fibres with a very reduced calibre resembling fetal fibres, such as those observed in group D of the present study, were described by Haltia et aL (1978) who also found a delay in their growth and differentiation. Timson and Dudenhoeffer (1985) also reported a reduction in the diameter and volume of muscle fibres, as well as a delay in the maturation and development of muscle fibres, as the most obvious alteration in malnutrition.

In the present study, the animals in group M showed an increase in the number of dark nuclei that seemed to be arranged in rows, together with changes in the muscle fibres in isolated and dispersed areas. These observations, as well as an increase in the number of nuclei in the sarcolemma (even though young individuals always have a greater number of nuclei), and an increase in the size of the endothelial nuclei, which were much more distinct, wider and paler than normal, were also reported by Vincent and Radermecker ([19-?]) in studies on the muscles of children with kwashiorkor.

The endomysium of the cells in group M was much thinner than in group N, suggesting alterations; however, the most striking sign was the thickness of the perimysium, which was greatly increased, which according to Oliveira (2006) is suggestive of oedema. Oedematous infiltration of the interfascicular septa and muscle atrophy were also reported by Vincent and Radermecker ([19-?]) who histologically investigated the deltoid and trapezius muscles of children with kwashiorkor. Trowell, Davies and Dean (1954), in addition to small interstitial masses of lymphocytes and macrophages, observed interstitial oedema with degeneration or vacuolisation of fat and thinning and atrophy of the muscle fibres in the heart muscle.

Based on the above information, it was observed that the greatest alterations were seen in the animals in group D, where in addition to enlarged perimysial spaces in some places these were filled with connective tissue which in certain regions infiltrated between isolated cells forming "islands" accompanied by macrophage-like cells and/or microblasts suggestive of an inflammatory process associated with delayed development. Similar observations were made by Durante (1902), who considered that pathological processes involving muscles resulted in an involution towards an embryonic state of the protoplasmic tissue, and that the proliferation of interstitial tissue was probably due to an irritative phenomenon caused by degenerative products.

The reduction in the calibre of the fibres with loss of cell contours and consequently irregular fibre contours observed in groups M and D in the present study corresponds to that reported by Nascimento (1990), who also states that in intrauterine protein deprivation there is a delay in the differentiation and development of the muscle fibres and the second motor neuron concomitantly in the interstitial tissue, which leads to diffuse masses of monocytes in the interfascicular spaces and occasionally also in the fascicles themselves. These observations support the previous reports by Durante (1902), who assumed that muscle hypoplasia and not atrophy occurs in malnutrition, as the tissue develops insufficiently from intrautero. This author reinforced that atrophy only occurs when there is a reduction in calibre in the case where the experimental animal was malnourished after its muscular system had matured.

According to Mendes and Waterlow (1957), the degree of muscle atrophy is best represented by the collagen content in the tissue. This author reports that the collagen content increases considerably during the diet, i.e. even in cases of protein deprivation, the collagen continues to grow at the same rate, which means that in these cases there is a greater amount of collagen visible in picrosirius stains in a similar way to that found in the animals in groups M and D, which showed a greater amount of collagen than the animals in group N in our study. These observations are reinforced by Oliveira (1999), who stated that early malnutrition would cause a delay in cell growth in relation to the development of connective tissue. It is very likely that in cases of malnutrition there is also a rapid and intense loss in the volume of contractile elements, as reported by Mayer (2008) in immobilised animals that showed muscle atrophy.

The polarised light assessment of the types of collagen fibres that make up the muscle envelopes showed a predominance of type I fibres in the animals in the nourished group (N), and type III fibres in the animals in groups M and D, where this collagen was clearly visible even in the endomysium, leading us to agree with Oliveira (2006) and conclude that the predominance of this type of fibre (type III) in the animals in groups D and M is probably indicative of a delay in the maturation of the collagen in the muscles of these animals.

When evaluating ultrastructural results, Racela et al. (1966) observed monkeys kept in a clinical condition of kwashiorkor, and observed muscle atrophy with a reduction in calibre and loss of myofilaments, enlarged and irregular mitochondria. Hansen-Smith, Maksud and Van Horn (1977), Oldfors and Sourander (1983, 1986), described in their studies mitochondrial loss in the muscle of rats subjected to protein malnutrition.

These findings support the results of the present study where in groups D and M we observed misalignment of the Z lines (fibres with low oxidative capacity), loss of myofilaments and few mitochondria. However, in group M we observed a loss of mitochondria, some enlarged and others forming rows between the myofibrils (fibres with high oxidative activity), as in the controls. These observations were also verified by Nascimento (1990), who also failed to observe changes in mitochondria in the malnourished animals, admitting that this may have been caused by artefacts in the technique and assuring that this preservation of mitochondria may have occurred due to the intensity and short period in which protein deprivation was carried out (15 and 30 day-old pups), similarly to our experimental groups, which were composed of 21 day-old malnourished pups. The results found in the literature regarding the ultrastructure of muscle fibres corroborate the findings of the present study, where it is possible to identify a greater length of band I in the animals from groups D and M, indicating that there was an increase in the length of the sarcomere and consequently a decrease in the number of sarcomeres.

CHAPTER 7

CONCLUSIONS

This research concludes that:

1. As for body weight, the animals in the Mozambique Group (M) had a higher weight at birth, but this decreased and became the same as the malnourished animals (D) at weaning (21 days);
2. In terms of size, the animals in group M had the same length as the malnourished animals (D) at birth and at weaning;
3. Qualitatively, the fibres of the animals in group M showed rounded fibres and enlarged nuclei, suggesting muscle atrophy;
4. Structurally, type III collagen fibres associated with muscle atrophy predominated in the M group animals;
5. The ultrastructure of the animals in group M showed misalignment of the Z lines and, in some places, rupture of the myofibrils;
6. In group M animals, glycolytic fibres had a lower percentage and ASTf was lower;
7. In the animals in group M, GO fibres had a lower proportion and lower AST, similar to that seen in group N;
8. The O fibres of the animals from Mozambique had a higher percentage and the same AST as group D, i.e. a smaller size.

REFERENCES

AGUAYO, V. M.; ROLEY, J. A.; MALANZELE, JMEERSHOEK, S. P. Opportunities for improving the quality of nutritional services in the national health system in Mozambique: findings from Manica Province. **Tropical Journal of Paediatrics,** v. 50, n. 5, p 314-318, 2004.

ALPER, C. M; MATTES, R. D. Effects of chronic peanut consumption on energy balance and hedonics. **International Journal of obesity.** 26:1129-1137, 2002.

Alves, P. H. M. **Histochemical characterisation of gastrocnemius muscle fibres in**

Wistar rats submitted to tenotomy and tenorrhaphy. 2010. 114 f. Dissertation (Master of Science)- Institute of Biomedical Sciences, University of São Paulo USP, São Paulo (2010).

ANDRADE, A.; PINTO, S.C.; OLIVEIRA, R.S. **Animais de laboratório: criação e experimentação.** Rio de Janeiro: Fiocruz, 2002. p. 275-279.

AZZARRI, C. (2010). Child Undernutrition in Mozambique. In: UNICEF SEMINAR, 2010, MAPUTO MINISTRY OF HEALTH, Mozambique.

BABLET, J.; NORMET, L. Les lesions histopathologiques de la bouffissure dÁnnam. **Bulletin of the Academy of Medicine,** v.117, p242. 1937.

BANDEIRA, S.O.; GASPAR, F.; PAGULA, F.P. African Ethnobotany and Healthcare: Emphasis on Mozambique, v. 39, p 70-73. 2001.

BANKER , B. Q.; ENGEL, A. G. Basic reactions of muscle. In: ENGEL A. G, FRANZINI-ARMSTRONG C, (Ed). **Myology.** New York: McGraw-Hill ,1994. p 832- 880.

BARNÁRD, R. J.; EDGERTON, V. R.; FURUKAWA, T.; PETER, J. B. Histochemical, biochemical, and contractile properties of red, white, and intermediate fibres. **The American Journal of Physiology,** v. 220, p. 410-416, 1971.

BEDI, K. S.; WARREN, M. A. The effects of undernutrition during early life on the rat optic nerve fibre number and size-frequency distribution. **The Journal of Comparative Neurology,** v. 219, n. 2, p 125-132, 1983.

BEHMER, O. A.; TOLOSA, E. M. C de.; FREITAS NETO, A. G. **Manual of techniques for normal and pathological histology.** São Paulo: EDART, 1975.

BOURNE, L.T.; HENDRICKS, M.K.; MARAIS, D.; ELEY, B. Addressing malnutrition in young children in South Africa. Setting the national context for paediatric food- based dietary guidelines. **Maternal Child Nutrition,** v. 3, n.4, p 230-238, 2007.

BRANDÃO, M. C. S. **Morphological and quantitative analysis of the myenteric plexus in the small intestine of young rats submitted to pre- and post-natal protein malnutrition.**
1998. Dissertation (Master of Science), Institute of Biomedical Sciences, University of São Paulo, São Paulo, 1998.

BROOKE, M. H.; KAISER, K. K. Muscle fibre types: How many and what kind? **Archives of Neurology,** v. 23, n. 4, p. 369-379, 1970.

BROOKE, M. H.; WILLIAMSON, E.; KAISER, K. K. The behaviour of four types in developing and reinnervated muscle. **Archives of Neurology,** v. 25, n. 4, p. 360- 366, 1971.

BURKE, R. E.; TSARIS, P. The correlation on physiological properties with histochemical characteristics in single muscle units. **Annals of the New York Academy of Sciences,** v. 228, p 145, 1974.

CARPENTER, S.; KARPATTI, G. General aspects of skeletal muscle. In: CARPENTER, S.; KARPATTI, G. **Pathology of skeletal muscle.** New York: Oxford University Press; 2001. p. 28-62.

CARVALHAES, M. A. B. L; BENÍCIO, M. H. D. Maternal capacity to care and child malnutrition. **Revista de Saúde Pública,** v 36, n. 2, p 188-197, 2002.

CASTELUCCI, P. (1999). **Morphoquantitative analysis of the myenteric plexus of the large intestine of rats submitted to pre- and post-natal protein malnutrition and post-natal renutrition.** 1999. Thesis (Doctorate in Sciences). Institute of Biomedical Sciences, University of São Paulo, São Paulo, 1999.

Electoral Census 2007. 22 August 2007, (AIM). Maputo, Mozambique.

CIENA, A. P.; LUQUES, I. U.; DIAS, F.J.; YOKOMIZO DE ALMEIDA, S. R.; IYOMASA, M.M.; WATANABE, LS. Ultrastructure of the myotendinous junction of the medial pterygoid muscle of adult and aged Wistar rats. **Micron,** v41, n. 8, p 1011- 1014, 2010.

COELHO, S. B.; SALES, R. L.; IYER, S. S.; BRESSAN, J.; COSTA, N. M. B.; LOKKO, P.; MATTES, R. Effects of peanut oil load on energy expenditure, body composition, lipid profile, and appetite in lean and overweight adults. **Nutrition,** v. 22, p. 585-592. 2006.

COLLING-SALTIN, A. S. Some quantitative biochemical evaluations of developing skeletal muscles in the human foetus. **Journal of the Neurological Sciences,** v. 39, n. 2-3, p. 187-198, 1978.

CONBOY, V. B.; SANTER, R. M.; SWIFT, G. L. Effects of prenatal undernutrition on prevertebral sympathetic neurons in the rat: a morphological and fluorescence histochemical study. **Jounal of Anatomy,** v. 154, p. 47-53, 1987.

CONRADI N. G.; SJÕSTRÕM, A.; KARLSSON, B.; SOURANDER, P. Functional development of the visual system in normal and protein deprived rats. II. Morphometric and biochemical studies on adult optic nerve. **Acta physiologica Scandinavica,** v. 125, n. 2, p. 277-83, 1985.

CONRADI, N. G.; SJÕSTRÕM, A.; RYDENHAG, B. Functional development of the visual system in normal and protein-deprived rats. VIL Postnatal development of optic nerve axons. **Acta physiologica Scandinavica,** v. 136, n. 4, p 597-603.1989.

CORNBLATH, D. R.; BROWN, M. J. Influence of malnutrition on developing rat peripheral nerves. **Experimental Neurology,** v. 99,n. 2, p. 403-411.1988.

D' ANTONA, G.; LANFRANCONI, F.; PELLEGRINO, M. A.; BROCCA, L.; ADAMI, R.; ROSSI, R.; MORO, G.; MIOTTI, D.; CANEPARI, M.; BOTTINELLI, R. Skeletal muscle hypertrophy and structure and function of skeletal muscle fibers in male body builders. **The Journal of Physiology,** v. 570, n. 3, p. 611-627, 2006.

DASTUR, D. K.; MANGHANI, D. K.; OSUNTOKUN, B. O.; SOURANDER, P.; KONDO, K. Neuromuscular and related changes in malnutrition: a review. **Jounal of Neurological Sciences,** v. 55, p 207. 1982.

DUBOWITZ, V. **Muscle biopsy:** a practical approach. London: Bailliére Tindall, 1985.

DURANTE, G. Anatomie pathologique des muscles. In: Cornil and Ranvier Manual d'Histologie Pathologique. Paris: Alcan, 1902.

EDGERTON, V. R.; SIMPSON, D. R. The Intermediate Muscle Fiber of Rats and Guinea Pigs**. The Journal of Histochemistry and Cytochemistry,** v. 1, p. 828-836, 1969.

ELDER, G. B.; KAKULAS, B. A. Hystochemical and contractile property changes during human muscle development. **Muscle Nerve,** v. 16, p. 1246-1253.1993.

ENESCO, M.; PUDDY, D. Increase in the number of nuclei and weight in skeletal muscle of rats of various ages. **American Journal of Anatomy,** v. 114, p. 235-244. 1964.

FAO/WHO. The State of food insecurity in the world. Eradicating world hunger-taking stock ten years after the world foot summit. 2006.

FARKAS-BARGETON, E.; DIEBLER, M. E.; ARSENIO-NUNES, M. L. WEHRLE, R.; ROSENBERG, B. Histochemical, quantitative and ultrastructural maturation of human foetal muscle. **Journal of the Neurological Sciences,** v. 31, n. 2, p. 43-49, 1977.

FERGUSON, E. L.; GIBSON, R.S.; OPARE-OBISAW, C.; OUNPUU, S.; THOMPSON, L.U.; LEHRFELD, J. The Zinc nutriture of preschool children living in two African Countries. **The Journal of Nutrition,** v. 123, p. 1487-1496.1993.

FINKEL, Y.; ROSENTHAL, P.; MOLLESTON, J.; ALONSO, E. The role of paediatric gastroenterology and nutrition in developing countries. **Journal of Paediatric Gastroenterology and Nutrition,** v. 39, p 445-447. 2004.

GIBSON, R.S.; YEUDALL, F.; DROST, N.; MTITIMUNI, B.M.; CULLINAN, T.R. Experiences of a community-based dietary intervention to enhance micronutrient adequacy of diets low in animal source foods and high in phytate: a case study in rural

Malawian children. **The Journal of Nutrition,** v. 133, p. 3992S-3999S. 2003.

GOPINATH, G.; ROY, S.; KARMARKAR, M.G. Effect of undernutrition and rehabilitation on the postnatal growth of skeletal muscle and nerve in the rat. **Indian Journal of Medical Research,** v. 77, p. 702-712.1983.

GREENE, E. C. Anatomy of the rat Muscles. In: GREENE, E. C. Transactions of the American Philosophical Society. New York: Hafner Publishing Co., v. 27, p. 59-83. 1959.

HAJ, A.J.; LEWIS, S.E.; GOLDSPINK, D.F.; MERRY, B.J.; HOLEHAN, A.M. The effect of chronic and acute dietary restriction on the growth and protein turnover of fast and slow types of rat skeletal muscle. **Comparative Biochemistry and Physiology,** v. 85, n. 2, p. 281-287, 1986.

HALTIA, M.; BERLIN, O.; SCHUCHT, H.; SOURANDER, P. Post-natal differentiation and growth of skeletal muscle fibers in normal and undernourished rats: a histochemical and morphometric study. **Journal of the neurological Sciences,** v. 36, p.25. 1978.

HÃMÃLÃINEN, N.; PETTE, D. The Histochemical Profiles of Fast Fiber Tupes IIB, IID and HA in Skeletal Muscles of Mouse, Rat and Rabbit. **The journal of histochemistry and cytochemistry,** v. 41, p. 733-743, 1993.

HAMPSHIRE, R.D.; AGUAYO, V.M.; HAROUNA, H.; ROLEY, J.A.; TARINI, A.; BAKER, S.K. Delivery of nutrition Services in health systems in sub-Saharan Africa: opportunities in Burkina Faso, Mozambique and Niger. **Public Health Nutrition** , v. 7, n. 8, p 1047-1053. 2004.

HANSEN-SMITH, F.M.; MAKSUD M.G.; VAN HORN, D.L. Effect of dietary protein restriction or food restriction on oxygen consumption and mitochondrial distribution in cardiac and red and white skeletal muscle of rats. **Journal of Nutrition,** v.107, p.525-533. 1977.

HEDLEY-WHITE, E. T.; MEUSER, C. S. The effect of undernutrition on myelination of rat sciatic nerve. **Laboratory investigation,** v 24, n. 2, p. 156-61.1971.

HEDLEY-WHITE, E.T. Myelination of rat sciatic nerve: comparision of undernutrition and cholesterol biosynthesis inhibition. **Jounal of Neuropathology & Experimental Neurology,** v. 32, n. 2. P. 284-302.1973.

HOTZ, C.; GIBSON, R.S. Traditional food-processing and preparation practices to enhance the bioavailability of micronutrients in plant-based diets. **The Jounal of Nutrition,** v.137, p. 1097-1100. 2007.

HOWARD, C. V.; REED, M. G. Three-dimensional measurement in microscopy. In: Unbiased stereology. Oxford: Bios Scientific Publishers, 2005. p 246.

HUGHES, S. M.; BLAU, H. M. Muscle fibre pattern is independent of cell lineage in postnatal rodent development. **Cell,** v. 68, p. 659-671. 1992.

JAHOOR, F.; BADALOO, A.; REID, M.; FORRESTER, T. Protein metabolism in severe childhood malnutrition. **An ls of Tropical Paediatrics,** v. 28, p. 87-101. 2008.

JUNQUEIRA, L. C. U.; BIGNOLAS, G.; BRENTANI, R. R. Picrosirius staining plus polarisation microscopy, a specific method for collagen detection in tissue sections. **The Histochemical Journal,** v.11, p. 447-455. 1979.

JUNQUEIRA, L.C.; CARNEIRO, J. **Basic histology.** Editora Guanabara Koogan S.A. 2004. pp 184-192.

KIRKMEYER , S. V; MATTES, R. D. Effect of food attributes on hunger and food intake. **International Journal of obesity,** v. 24, p. 1167-1175. 2000.

LINDBOE, C.F.; PRESTHUS, J. Effects of denervation, immobilisation and cachexia on fibre size in the anterior tibial muscle of the rat. Acta Neuropathologica, v. 66, p.42-51, 1985.

LINDBOE, CH. F.; TORVIK, A. The effects of ageing, cachexia and neoplasms on striated muscle- quantitative histological and histochemical observations on na autopsy material, Acta Neuropathologica, (Berl.) v. 57, p. 85-92. 1982.

LOW, J.W.; ARIMOND, M.; OSMAN, N.; CUNGURA, B.; ZANO, F.; TSCHIRLEY, D. A food-based approach introducing orange-fleshed sweet potatoes increased Vitamin A intake and serum retinol concentrations in young children in rural Mozambique. **The Journal of Nutrition,** v. 137, p. 1320-1327. 2007.

LUNET, N.; MURANGY, Â.; BRANCO, A.C.; CALANE, A.; BUCAR, A.; ELYAS, B.; BRANQUINHO, B.; NAMBURETE, D.; BARROS, H. Malnutrition in children in Maputo, Mozambique. **Revista de Epidemiologia. Arquivos de Medicina,** v. 16, n. 6, p. 26-29. 2002.

MAKI, K.; CAMPOS, S.D. Experimental malnutrition: anatomopathological and biochemical results of the administration of hypoprotein diets to young albino rats. **Memórias do Instituto Oswaldo Cruz,** v. 73, p. 153-181. 1975.

MANGER, M.S.; MCKENZIE, J.E.; WINICHAGOON, P.; GRAY, A.; CHAVASIT, V.; PONGCHAROEN, T.; GOWACHIRAPANT, S.; RYAN, B.; WASANTWISUT, E.; GIBSON, R. A micronutrient-fortified seasoning powder reduces morbidity and improves short-term cognitive function, but has no effect on anthropometric measures in primary school children in northeast Thailand: a randomized controlled trial. 2008.

MARTINS, H. Causes, consequences and costs of malnutrition in Mozambique. Instituto de Higiene e Medicina Tropical de Lisboa, Lisbon, Portugal. 2008.

MATHIENSEN, A. L. **Morphological and quantitative study of the different types of fibres in the biceps brachii muscle of wistar rats submitted to protein malnutrition.** 2004. Dissertation (Master's Degree in Morphofunctional Sciences), Institute of Biomedical Sciences, University of São Paulo. 2004

MAYER, W. P. **Structural, ultrastructural and morphoquantitative characteristics of the fibres of the tibialis anterior and soleus muscles of young rats submitted to immobilisation of the talocrural joint.** 2008. 117f. Dissertation (Master's Degree), Institute of Biomedical Sciences, University of São Paulo, São Paulo, 2008.

MISAU. Ministry of Health (MISAU). National Health Directorate. Nutrition Department. Mozambique: investing in nutrition is reducing poverty. Analysis of nutritional problems in children and women. Mozambique, 2002.

MONTEIRO, C. A. The dimension of poverty, malnutrition and hunger in Brazil. **Estudos avançados,** v. 17, n. 48, p. 7-20. 2003.

MONTEIRO, C. A. The determinants of child malnutrition in the Ribeira Valley. **Cadernos de pesquisa,** v. 29, p. 57-75. 1979.

MONTGOMERY, R.D. Muscle morphology in infantile protein malnutrition. **Journal of clinical Pathology,** v. 15, p. 511-521.1962.

MOORE, K. L.jPERSAUD, T.V.N. **Embriologia clínica-** 8ª Edição, 2008. p 364-366.

NASCIMENTO, O. J. M.; MADI, K.; SILVA, J. B. G.; SOARES FILHO, J.; HAHN, M. D.; COUTO., B.; FREITAS, M. R. G. Considerations on striated muscle in protein malnutrition. **Archives of Neuro-Psychiatry,** v. 48, p. 395-402. 1990.

NÓBREGA, F.J. **Intrauterine and postnatal malnutrition.** SÃO PAULO: PANAMED EDITORIAL, 1986.

OGATA, T. A. Histochemical studies on red and white muscle fibres. Part III. Activity of the diphosphopyridine nucleotide diaphorase and triphosphopyridine nucleotide diaphorase in muscle fibres. **Acta Medica Okayama,** v. 12, p. 233-240, 1958

OLDFORDS, A.; PERSSON, M. Peripheral nerve fibre degeneration in protein- deprived young rats. An ultrastructural study. **Acta neuropathologica** (Berl.), v. 57, n. 1, p. 1-6. 1982.

OLDFORDS, A.; SOURANDER, P. Nutritional Rehabilition of skeletal muscle in protein-deprived Young rats. **Journal of the Neurological Sciences,** v. 75, p.173- 179. 1986.

OLDFORS, A.; SOURANDER, P. Muscle changes in protein-deprived young rats-A morphometrical, histochemical and ultrastructural study. **Journal of the Neurological Sciences,** v.599, p.291-302. 1983.

OLIVEIRA, F. Histochemical characteristics of the muscle fibres of the gastrocnemius muscle of Wistar rats subjected to thermal injury by scalding. 2006. 140 f. Thesis (Doctorate), Institute of Biomedical Sciences, University of São Paulo, São Paulo, 2006.

OLIVEIRA, F.L.C.; OLIVEIRA, A.S.B.; SCHMIDT, B.,;AMANCIO, OM.S. Intrauterine energy malnutrition in rats: musculoskeletal changes in the 1st[a] and 2nd[a] generations. **Jornal de Pediatria,** v. 75, n. 5, p. 350-356.1999.

PACHECO, M. E. Structure and typification of the skeletal muscle fibres of the medial gastrocnemius muscle of rats in malnutrition and early corrected renutrition. 2006. Dissertation (Master's Degree in Morphofunctional Sciences), Institute of Biomedical Sciences, University of São Paulo, São Paulo. 2006.

PETER, J.; BARNARD, R.; EDGERTON, V. Metabolic profiles of three fibre types of skeletal muscle in guinea pigs and rabbits. **Biochemistry,** v. 11, p. 2627- 2634, 1972.

PETTE, D; STARON, R. S. Transitions of muscle fibre phenotypic profiles. **Histochemistry and Cell Biology,** v. 115, p. 359-372, 2001.

PRISTA, A., MAIA; J.A.R.; DAMASCENO, A.; BEUNEN, G. Anthropometric indicators of nutritional status: implications for fitness, activity, and health in school- age children and adolescents from Maputo, Mozambique. **The American Journal of Clinical Nutrition ,** v. 77, p. 952-959. 2003.

RACELA, A. S.; GRADY, H. J.; HIGGINSON, J.; SVOBODA, D. Protein deficiency in Rhesus Monkeys. University of Kansas Medical Centre, Kansas City, Kansas. v.49, n. 3, p. 419-443. 1966.

RANA, S. V.; CHOPRA, J. S.; MEHTA, S.; CHAND, U. K.; SHARMA, U. Electrophysiological and histopathological changes in peripheral nerves of postnatally undernourished and rehabilitated Young rhesus monkeys. **Indian Jounal of Experimental Biology,**v. 29, n. 4, 385-390. 1991.

REEVES, P. G.; NIELSEN, F. H.; FAHEY G. C. Jr. AIN-93 purified diets for laboratory rodents: final report of the American Institute of Nutrition ad hoc writing committee on the reformulation of the AIN-76A rodent diet. **Journal of Nutrition,** v. 123, n. 11.p. 1939-51.1993.

PRELIMINARY REPORT MOZAMBIQUE: Demographic and Health Survey, Ministry of Health, Maputo, Mozambique, 2011

REYNOLDS, E. S. The use of lead citrate at high pH as an electron opaque stain in electron microscopy. **The Journal of Cell Biology,** v. 18, p. 208-213, 1963.

ROSE, D.; TSCHIRLEY, D. Predicting dietary intakes with simple food recall information: a case study from the rural Mozambique. **European Journal of Clinical Nutrition,** v. 57, p. 1212-1221. 2003.

ROSE, S.J.; ROTHSTEIN, J. M. Muscle Mutability. Part 1. General Concepts and adaptations to Altered Patterns of Use. **Physical Therapy,** v. 62, n.12, p. 1773-1787, 1982.

ROSS, M. H.; PAWLINA, W. **Histology. Text and atlas.** 5ª Edition, 2006. p 283-288.

ROY, S.; SINGH, N.; DEO, M. G.; RAMALINGASWAMI, V. Ultrastructure of skeletal muscle and peripheral nerve in experimental protein deficiency and its correlation with nerve conduction studies. **Journal of the neurological Sciences,** v. 17, n.4, p. 399-409.1972.

SACHDEV, K. K.; TAORI, G. M.; PEREIRA, S. M. Neuromuscular status in protein-calorie malnutrition. **Neurology,**v. 21, p. 801.1971.

SAHN, D. E.; ALDERMAN, H. On the Determinants of nutrition in Mozambique: the importance of age-specific effects. **World Development,** v. 25, n. 4, p. 577-588. 1997.

SALLES, R. L.; COSTA, N. M. B.; MONTEIRO, J. B. R.; PELUZIO, M. C. G.; COELHO, S. B.; OLIVEIRA, C. G.; MATTES, R. Effects of peanut, saffron and olive oil on body composition, energy metabolism, lipid profile and food intake of normolipidemic eutrophic individuals. **Revista de nutrição,** v.18, n.4, p. 499-511.2005.

SARNAT, H. B. Developmental disorders of muscle. In: Mastaglia, F. L.; Walton, J. Skeletal muscle pathology. New York: Churchill Livingstone, 1982. p. 140-160.

TECHNICAL SECRETARIAT FOR FOOD SAFETY AND NUTRITION (SETSAN). Ministry of Agriculture and Rural Development, National Directorate of Agriculture. 2005. Case study on food and nutrition security in Mozambique's development process.

SHER, J.; CARDASIS, C. Skeletal muscle fibre types in the adult mouse. **Acta Neurológica Scandinavica,** v. 54, p. 46-56, 1976.

SIECK, G.C.; LEWIS, M.L; BLANCO, C.E. Effects of undernutrition on diaphragm fibre size, SDH activity, and fatigue resistance. **Jounal of Applied Physiology,** v. 66, n. 55, p. 2196-2205.1989.

SILVA, R. H. G. Electromyographic and morphological analysis of the masseter muscles of capuchin monkeys (Cebus apella) submitted to unilateral section of the masseteric nerve. 1999.120 f. Thesis (Doctorate) - Araçatuba School of Dentistry, São Paulo State University, São Paulo.

SIMA, A.; SOURANDER, P. The effect of early undernutrition on the caliber spectrum of the rat optic nerve. A morphometric ultrastructural study. **Acta neuropathologica** (BerL), v. 28, n. 2, p. 151-60.1974.

SIMONEAU, J. A.; BOUCHARD, C. Genetic determinism of fibre type proportion inhuman skeletal muscle. **The FASEB journal,** v. 9, p. 1091-1095, 1995.

STANDRING, S., BORLEY, N., COLLINS, P., CROSSMAN, A. R., GATZOULIS, M., HEALY, J. C., JOHNSON, D., MAHADEVAN, V., NEWELL, R. L. M" WIGLEY, C. B. (ed). 2008. Gray's Anatomy. The anatomical Basis of clinical practice. p 103-108.

STARON, R.S.; KRAEMER, W.J.; HIKIDA, R.S.; FRY, A.C.; MURRAY, D.; CAMPOS, G.E. Fiber type composition of four hindlimb muscles of adult Fisher 344 rats. **Histochemistry and Cell Biology,** v.111, n.2. p. 117-123, 1999.

TIMSON, B.F.; DUDENHOEFFER, G.A. The effect of severe dietary protein restriction on skeletal muscle fibre number, area and composition in weanling rats. **Journal of Animal Science,** v. 61, p.416-22.1985.

TROWELL, H. C.; DAVIES, J. N. P.; DEAN, R. F. A.. Kwashiorkor. London, Arnold, 1954.p. 153-195.

VAN BOGAERT, L.; RADERMECKER, M. A.; JANSSEN, P.; GATAERA, F. The muscular lesions of Kwashiorkor compared with those of other oedematous conditions and malnutrition in the Congolese infant. **Acta Neuropathologica** (BerL), v. 1, p. 363-383.1962.

VINCENT, M.; RADERMECKER, M-A. ([19-?]). Histological investigation of muscle tissue in kwashiorkor. p.511-517.

WAINIO, W. W.; ALLISON, J. B.; KREMZNER, L. T.; BERNSTEIN, E.; ARONOFF, M. Enzymes in protein depletion: III. Enzymes of brain, kidney, skeletal muscle and spleen. **Journal of Nutrition,** v. 67, p. 197. 1959.

WALSH, G.; DE VIVO, D.; OLSON, W. Histochemical and ultrastructural changes in rat muscle. **Archives of neurology,** v. 24, p. 83. 1971.

WHITTEN, W.K. (1956). Modification of the oestrous cycle of the mouse by external stimuli associated with the male. **Journal of Endocrinology,** v. 13, n.4, p. 399-404.

WILLIAMS, J.P.; HUGHES, P.C. Muscle growth during neonatal undernutrition and subsequent rehabilitation in the rat. **Acta Anatomica** (Basel), v. 101, n.3, p. 249-254. 1978.

WINICHAGOON, P.; MCKENZIE, J.E.; CHAVASIT, V.; PONGCHAROEN, T.; GOWACHIRAPANT, S.; BOONPRADERM, A.; MANGER, M.S.; BAILEY, K.B.; WASANTWISUT, E.; GIBSON, R. A micronutrient fortified seasoning powder enhances the haemoglobin, Zinc, and lodine status of primary school children in North East Thailand: A randomized controlled trial of efficacy. **The Journal of Nutrition,** v. 136, p. 1617-1623. 2006.

WINICK, M.; NOBLE, A. Cellular response in rats during malnutrition at various ages. **Journal of Nutrition,** v. 89, n. 3.p. 300-306.1966.

WORLD HEALTH ORGANISATION (WHO). Bulletin of World Health Organisation, v.75, n. 1, p.81-93. 1997

WORLD HEALTH ORGANISATION (WHO) Bulletin of World Health Organisation, v.80 n.3. 2002.

WORLD HEALTH ORGANISATION (WHO). The World Health Report. Make every mother and child count. 2005.

ZEMAN, F.J. Effect of the young rat of maternal protein restriction. **Journal of Nutrition,** v. 93, n. 2, p. 167-73.1967.
WEBSITES CONSULTED

www.unicef.org/mozambique

http://www.pastoraldacrianca.org.br

ANNEX A: **TABLE 1: NUTRITIONAL STATUS OF CHILDREN IN MOZAMBIQUE BY PROVINCE** (extracted from: Mozambique Preliminary Report: Demographic and Health Survey, 2011).

Quadro 11 Estado nutricional das crianças

Percentagem de crianças menores de cinco anos classificadas como malnutridas, segundo três indices antropométricas do estado nutricional: Altura por idade, peso por altura, e peso por idade segundo características sociodemográficas, Moçambique 2011

	Altura por idade[1]			Peso por altura				Peso por idade				
Características sociodemográficas	Percentagem abaixo de -3 DP	Percentagem abaixo de -2 DP[2]	Valor médio de Z (DP)	Percentagem abaixo de -3 DP	Percentagem abaixo de -2 DP[2]	Percentagem abaixo de +2 DP	Valor médio de Z (DP)	Percentagem abaixo de -3 DP	Percentagem abaixo de -2 DP[2]	Percentagem abaixo de +2 DP	Valor médio de Z (DP)	Número de crianças
Idade em meses												
<6	12.8	27.6	(1.1)	4.3	10.5	18.0	0.3	6.8	15.3	2.0	(0.7)	1,012
6-8	13.6	29.1	(1.0)	2.2	10.2	8.7	0.0	4.6	16.6	1.6	(0.7)	538
9-11	16.6	34.6	(1.3)	2.7	9.1	7.4	(0.1)	6.3	17.5	1.5	(0.8)	604
12-17	18.6	44.0	(1.6)	4.5	10.1	7.3	(0.1)	6.6	16.7	0.9	(0.9)	1,192
18-23	26.0	48.1	(1.9)	2.4	6.4	6.1	0.1	3.6	16.4	0.6	(0.9)	969
24-35	22.9	49.3	(1.9)	2.2	5.3	7.6	0.3	3.5	13.6	0.5	(0.8)	1,954
36-47	22.8	47.0	(1.9)	0.5	2.1	5.0	0.3	3.3	13.6	0.3	(0.9)	2,095
48-59	17.0	41.6	(1.8)	1.0	3.3	4.5	0.2	2.1	14.5	0.1	(1.0)	1,949
Sexo												
Masculino	21.5	44.7	(1.8)	2.5	6.4	7.5	0.2	5.0	16.6	0.7	(0.9)	5,140
Feminino	17.9	40.5	(1.6)	1.8	5.4	7.2	0.2	3.2	13.3	0.6	(0.8)	5,173
Área de residência												
Urbana	15.4	35.0	(1.4)	1.4	3.8	8.2	0.3	2.7	9.8	1.5	(0.6)	2,859
Rural	21.3	45.5	(1.8)	2.4	6.7	7.1	0.1	4.6	16.9	0.4	(1.0)	7,455
Província												
Niassa	24.1	46.8	(1.8)	1.3	3.8	6.9	0.2	5.1	18.2	1.0	(0.9)	634
Cabo Delgado	26.7	52.7	(2.1)	1.5	5.6	6.3	0.1	6.4	20.6	0.4	(1.1)	879
Nampula	30.0	55.3	(2.1)	3.0	6.5	8.9	0.3	4.5	15.5	0.2	(1.0)	1,560
Zambézia	21.0	45.2	(1.7)	4.2	9.4	4.6	(0.2)	5.9	21.3	0.5	(1.1)	2,080
Tete	19.3	44.2	(1.7)	1.7	5.6	6.7	0.1	4.1	17.0	0.2	(0.9)	1,342
Manica	18.2	41.9	(1.6)	2.5	6.7	10.3	0.2	2.5	10.8	0.8	(0.8)	671
Sofala	14.8	35.7	(1.4)	1.6	7.4	7.4	0.1	3.2	11.3	0.9	(0.8)	1,082
Inhambane	15.8	36.0	(1.6)	0.5	2.2	10.8	0.6	1.2	6.9	1.4	(0.5)	576
Gaza	8.5	26.8	(1.3)	0.3	1.0	7.0	0.5	1.8	6.3	1.4	(0.4)	559
Maputo província	6.0	22.7	(1.2)	0.5	2.1	9.5	0.5	1.7	7.4	0.9	(0.3)	576
Maputo cidade	7.4	23.2	(1.0)	0.6	2.2	8.6	0.6	2.4	5.4	2.8	(0.2)	354

MIX
Papier aus verantwortungsvollen Quellen
Paper from responsible sources
FSC® C105338

Printed by Books on Demand GmbH, Norderstedt / Germany